Cities and Nature

Series Editors

Peter Newman, Sustainability Policy Institute, Curtin University, Perth, WA, Australia
Cheryl Desha, School of Engineering and Built Environment, Griffith University, Nathan, QLD, Australia

Cities and Nature fosters high-quality multi-disciplinary research addressing the interface between cities and the natural environment. It provides a valuable source of relevant knowledge for researchers, planners and policy-makers. The series welcomes empirically based, cutting-edge and theoretical research in urban geography, urban planning, environmental planning, urban ecology, regional science and economics. It publishes peer-reviewed edited and authored volumes on topics dealing with the urban and the environment nexus, including: spatial dynamics of urban built areas, urban and peri-urban agriculture, urban greening and green infrastructure, environmental planning, urban forests, urban ecology, regional dynamics and landscape fragmentation.

More information about this series at http://www.springer.com/series/10068

Jana Söderlund

The Emergence of Biophilic Design

Springer

Jana Söderlund
Sustainability Policy Institute
Curtin University
Perth, WA, Australia

ISSN 2520-8306 ISSN 2520-8314 (electronic)
Cities and Nature
ISBN 978-3-030-29812-8 ISBN 978-3-030-29813-5 (eBook)
https://doi.org/10.1007/978-3-030-29813-5

This Springer imprint is published by the registered company Springer Nature Switzerland AG
The registered company address is: Gewerbestrasse 11, 6330 Cham, Switzerland

To people with passion
Passion is the fuel for action
and to
Jett and Bodhi

Foreword

Forward: Jana's Quest

Jana Söderlund has been on a quest to bring more nature into our homes and communities. In 2013, she travelled to my home city, Charlottesville, Virginia, to participate in what became the launch of a new global Biophilic Cities Network. She presented her work in helping the Australian city of Fremantle (which is now itself a partner city in the global network) develop and implement an innovative program for installing and studying small green walls around town. It was a program called Green Skins and at our launch event, Jana led a hands-on workshop designing and building green walls. Her passion and commitment to biophilic design and planning were clearly evident then. These are things she wears on her sleeves.

During the conference, one of our keynote speakers, Stephen Kellert, an inspiration to many of us, discussed the important role of beauty. He cited survey data that Jana had collected about reactions people had to the green walls installed in Fremantle. Their many ecological benefits are important, to be sure, but Jana had found that the aesthetic qualities, the beauty that residents experienced when they saw them, ended up being the most important factor in their appeal. It was (and is) an important insight: we need nature in cities, in many forms and for many reasons, but it is often the essential joy and delight that they offer that carries the day.

Most impressively, following the Biophilic Cities launch, Jana embarked on a remarkable journey around the US to meet with leaders and the "pioneers", and to see firsthand the leading biophilic projects and cities, and essentially to document the emergence and progress of this remarkable global movement. A deep curiosity and a dogged spirit propelled her forward. Accompanied by supportive son Jett, Jana covered a remarkable amount of distance in a short period of time to capture these stories.

The result is the book before you: an impressive recording of the many inspiring leading ideas, people and places that make up (at least in this genesis phase of) the biophilic design movement. It is at once a comprehensive and creative melding of interviews and field visits, to create a compelling account of this inspiring movement. It is a bit of hope when we need it the most and will likely become a standard reference for those in the future seeking to understand what was percolating and why during this especially formative period.

Thank you Jana!

—Tim Beatley

Biophilia, as it appeared in the design world sparked by E. O. Wilson in *Biophilia* (1984) and expanded in *The Biophilia Hypothesis* (Kellert and Wilson, 1993), focused on humanity's evolved need for connection to the natural world. The recognition of this desire to affiliate with nature created the field of biophilic design in the built environment. It grew from a relatively small cadre of designers and researchers into an expansive endeavour touching many fields of design—from buildings and interior spaces to urban design with the stated goal of enhancing human health and happiness through ongoing connection to the natural world.

A neglected topic in the rapid growth of biophilic design is the role of human connectedness. As Jana Söderlund argues in *The Emergence of Biophilic Design*, biophilia is not just a design issue. It is also a social movement built around the idea that connection to nature is a basic human need. It is the recognition of this need that captured the attention of so many people—not just designers.

Addressing the social underpinnings of biophilic design raises important new issues including the democratisation of biophilia. If connection to nature is, indeed, an evolved human need, then it is a need shared by all—not just those who can afford to live in areas with green space and work in buildings with connections to natural features and elements. An extensive body of research shows that nature has social as well as emotional and physiological benefits.

The broad application of biophilic design, especially in cities where most people now live, is more possible now as it gains wide appeal. Jana Söderlund's book opens our eyes to the importance of networks of people pursuing this shared vision for connection to nature in everyday life.

—Judith Heerwagen

"…enjoyment of scenery employs the mind without fatigue and yet exercises it, tranquilizes it and yet enlivens it; and thus through the influence of the mind over the body, gives the effect of refreshing rest and reinvigoration to the whole system."

Frederick Law Olmsted, 1865, *An Introduction to Yosemite and Mariposa Grove: A Preliminary Report*

In his letter to the US Congress, on the design for the proposed Yosemite and Mariposa Grove National Parks, Olmsted was describing the psychological and physiological benefits of experiencing nature. His observation about these responses later becomes known as Attention Restoration Theory. In recent years, neuroscientific studies have confirmed Olmsted's insights—experiences of nature do lower stress and change the way the brain operates. These are measured outcomes of humans connecting with nature, the core of the concept of biophilia. The biophilic design movement seeks to bring those benefits to everyday urban experiences.

In addition to lowered stress, improved cognitive function and general well-being, there is some emerging evidence that experiences of nature encourage more prosocial behaviours, make us more charitable. With the growing crisis of climate change and the loss of other species, the hope is that biophilic experiences will change attitudes and political behaviours.

The realisation that humans have now become predominately an urban species, means the need to reconnect with the qualities of the natural environment in which we evolved are becoming ever more important. Parks, gardens, water features and nature views have long been evident in the precincts of the wealthy. That we place value on just being able to see these pieces of nature can be seen in real estate prices, and even in room rates for hotels. Today we must extend those experiences to everyone, every day.

This is a lofty, ambitious and necessary goal. It will require us to remake our buildings and cities, and it must involve more than just a handful of committed designers. Bringing nature into the daily experience of urban dwellers will require designers, planners, owners, city officials, citizens—nothing short of a social movement.

—Bill Browning

Charlottesville, USA — Tim Beatley
Seattle, USA — Judith Heerwagen
New York, USA — Bill Browning

References

Kellert SR, Wilson EO (eds) (1993) The biophilia hypothesis. Island Press, Washington, DC
Wilson EO (1984) Biophilia. Harvard University Press, Massachusetts

Preface

Cities around the world are growing dramatically. More people now, in 2019, live in cities than in rural areas (Lehmann 2015, p. 1). "By 2030, 60 percent of the world population, or 4.9 billion people, are expected to live in urban areas" (Girardet 2015, p. 4). Human settlement has not occurred in such a way before. High fossil fuel and resource consumption has enabled this expansive urbanisation while contributing significantly to global warming and climate change (Girardet 2015; Lehmann 2015; Olsson and Haas 2014). Yet, cities are not only expanding, they are changing in their roles and in their function. Deindustrialisation, increased mobility and a growing service sector have seen urban areas transform into post-industrial knowledge-based economies of consumption rather than production (Olsson and Haas 2014). Emerging from this shift in focus of cities' function is an evolving change in form: a change in the way buildings are being designed, constructed and landscaped.

Typically, industrialised cities, with their focus on function, became harsh, engineered landscapes of paved surfaces and inner-city urban canyons. Psychoanalyst Fromm (1964), and more recently Salingaros and Masden (2008), recognised this, proposing that contemporary cities can be viewed as mechanistic, sterile, industrialised, commoditised and devoid of nature. Fromm (1964) also argued that urban dwellers were facing a disconnect from nature and loss of the psychological benefits that can ensue from a healthy human–nature relationship. To follow a positive, progressive pathway in life Fromm proposed that a love of life was necessary. He coined the term 'biophilia' to express this human–nature connection, with 'bio' meaning life and 'philia', the opposite of 'phobia', meaning attraction or love (Fromm 1964).

Significantly, years later another scholar, prominent sociobiologist, Edward Wilson, utilised Fromm's term biophilia to describe emotions which were provoked in a period of immersion in nature (Wilson 1984). Wilson defined biophilia as the "innate tendency to focus on life and life-like processes" (Wilson 1984, p. 1). His book, *Biophilia*, presented a similar perspective to earlier conservationists such as Naess (1989), though with a perception of the human connection to nature as an innate, biological need, not solely an inherent interdependence. Traditionally there have always been thinkers who have recognised human connection and interdependence with nature and encouraged others to do the same (Leopold 1949; Naess 1989). Illustrative descriptions of nature and ecosystems were set in the forests, rivers and natural areas, not in the cities. People travelled outside of the city to have a nature experience. Conservation, environmental or deep ecology movements tended to be anti cities, focused on protecting the nature that remained outside urban areas. Yet the shift in the function of cities, globalisation and the emergence of compact city theory (Lehmann 2015) is paving the way for a redefining of urbanites' relationship with nature, utilising the term biophilia introduced by Fromm in the 1960s.

With technological advancements supported by academic research and literature, the inclusion of nature in cities has, in many cities, been rapidly, and globally, expanding. What began with a term coined by Fromm has attracted further investigation and development by interested people with common goals and a desire to enable a greater opportunity for urban dwellers to affiliate with nature, and all the benefits this provides, within the built environment. The focus on the human–nature connection is no longer relegated to conservationists and natural areas outside of cities; it is coming from urban inhabitants. A social movement based on biophilic design has evolved.

The movement appears to be supported by increasing urban population and changing city function which has led to a mutable dynamic and interplay between urban places and spaces. This recent and expanding transformation in human urban settlement is requiring a new approach to building cities. Cities need to be designed, planned, built and retrofitted to be sustainable and liveable (Storey and Kang 2015). Higher building density, urban canyons and paved surfaces modify local climate, particularly temperature, leading to a phenomenon known as the urban heat island effect (Mills 2015). This correlation between increasing global urban population, climate change and urban heat island effect, and the need for liveable, higher density cities is repeated throughout sustainability literature discussing cities and design (Newman and Jennings 2008; Owen 2009; Steiner 2011). Within this framework, nature and biophilic design are finding a renewed status and recognition as essential components of a healthy, sustainable city (Lehmann 2015, p. 20).

Global examples of biophilic design demonstrate that in many instances the initiative is not purely a functional response to a city's sustainability challenges. There is a motivation beyond the function. Indicators are there that a shift in the approach to the human–nature urban connection has occurred. The principles of biophilic design represent these newly emerging initiatives that are occurring in cities. The origins of biophilic design and the fact that it signifies a social shift in thinking contribute towards defining it as a social movement.

This book is the result of a Ph.D. focused on the emerging social movement of biophilic design. It was undertaken within a framework of heuristic inquiry which required methodologies of an immersive journey after initial engagement with the focus of inquiry. This is the story of part of the initial research and engagement, followed by immersion in a journey of connection, discovery and learning.

The book primarily consists of conversations in interviews held with the pioneers of biophilic design initiatives throughout North America and two in my home town of Fremantle Western Australia. These stories contain a richness of information, not just about biophilic design, but about people, their passions and their intuitive knowing.

The motivators and drivers of the biophilic design movement were revealed in these interviews and contributed towards a framework for mainstreaming biophilic design initiatives. Interviewees were also asked about their connection to nature and for one word to describe how they felt when immersed in nature and a word cloud revealed the feeling that was most common.

The Term Biophilic Design

Throughout the immersive journey of meetings and interviews, there was significant discussion of the word biophilia and the term biophilic design. It can be seen as a new-fangled term for what some perceive they have already been doing, or sound like a fad. Some landscape architects and architects have embraced the term biophilic design, appreciating that it encompasses more than traditional landscape architecture or architecture by including building-integrated vegetation and the spaces, patterns and materials of nature in building design. They have discovered a term for what they intuitively design with new understandings, rationale and perceptions. Some professionals considered that it can explain concepts to others beyond landscape architects, especially how nature is integrated into buildings. 'Biophilic' can provide a context and a broader framework for including the social quality of life aspects.

In discussions with academics over the term biophilic design some said their preference would be to use another term, yet no one could think of a replacement that expressed the intrinsic nature connection and the human–nature benefits that result. There are terms that convey the green nature aspect such as living architecture, vertical greenery, skyrise greenery, green facades, living walls and so on. None also convey the human component.

There was agreement that something was needed: a term that would unite the fragmented disciplines and the fragmented research; a term that would provide a framework and context for great urban design and planning; a term that illustrates the change in conservation and the human–nature connection in cities. By providing the biophilic design broader framework, research into the design principles or the derived benefits could unite under a single term. This would facilitate researching and conveying the multiple benefits.

Undertaking this research reinforced what a large umbrella biophilic design is and how the parameters move between disciplines and thinkers. Should biophilic design include all aspects of nature in the city with the focus on the shift in the paradigm surrounding it? Does the term need to be narrowed to just include the architectural innovations and biophilic understanding of the built environment?

Since completion of the initial research, biophilic design has become mainstream in many professions and global discussions. While the terms usage has increased rapidly in the last years, in some areas it is attributed just to the greenery of nature and there lacks a full appreciation of the multiple attributes and benefits. Education and understanding is increasing as biophilic design becomes an accepted profession.

The Significance of this Book

With the current literature focussing on the why and how of biophilic design, with successful models and examples outlined throughout, little has been written on the social movement that the growing literature, media and personal experience indicates is occurring.

Beatley (2011, p. 66, 150) suggests the need to build social capital and leadership in biophilic urbanism, by gaining a better understanding of the social aspects involved. Van den Born et al. (2001) researched people's visions of nature, assessing their biophilia (nature-friendliness), involving both qualitative and quantitative methodology. Their summary suggested further lines of research including how more could be done to "...discover lay people's visions, and bring improved articulation of these concepts back to their originators" (Van den Born et al. 2001). They go on to suggest that "the second and quite policy relevant line for further research concerns the origins of people's visions of nature" (Van den Born et al. 2001). Their research had revealed a childhood/adulthood linkage with people's connection and view of nature, and they expressed concern that lack of nature for contemporary children was not perpetuating this connection and the "biophilic cohort" could die off (Van den Born et al. 2001). For them it is critical to see biophilic design in cities and to gain an understanding of the social drivers and motivators in achieving this.

Examination of the benefits to humans and the urban environment that biophilic design can bring raises the question of why cities have been built with such exclusion of nature. In Hillary Brown's prologue to *Biophilic Design* (Brown 2008, p. xiv), she also queries this, viewing "this design sensibility" as "one more intuitively biologic in nature", so much so that it should be an "inherent organising principle of all works of architecture". However, common sense doesn't mean it is actually part of professional practice and policy; it has to be driven by the broader culture. This book can assist in addressing these points by understanding the "drivers of the broader culture".

The documenting of the emergence of the social movement of biophilic design not only provides insights into biophilic design but also contributes to the understanding of contemporary social movements and their significance and potential to dramatically impact and change societies. This book tells the story of an idea, a theory, and how this theory developed into a concept which currently is of global significance in its implementation and which shows potential for embedding in the way cities are designed.

In identifying the motivators and drivers pushing the developmental social movement stages, the research provides a template of knowledge with the potential to assist further global development of biophilic design.

Government, non-government organisations, industry involved with biophilic initiatives and academia may gain deeper perspectives on how to progress biophilic design in their cities by understanding what may motivate and drive the players in different arenas to embrace the concept. By examining the characteristics of the biophilic design movement and identifying the strengths, characteristics, power flows, durability and triggers, policymakers will have a foundation on which to base decision-making in implementing new planning guidelines for urban design. Identifying biophilic design as a significant social movement will assist in the growth of the concept in cities. It will provide guidelines on the potential attributes to seek out and nurture, and it will provide security for people wishing to invest in biophilic businesses or implementation measures.

Perth, Australia Jana Söderlund

References

Beatley T (2011) Biophilic cities: integrating nature into urban design and planning. Island Press, Washington, DC

Brown H (2008) Prologue: in retrospect. In: Kellert SR, Heerwagen J, Mador M (eds) Biophilic design: the theory, science, and practice of bringing buildings to life. Wiley, Hoboken, NJ

Fromm E (1964) The heart of man. Harper and Row, New York

Girardet H (2015) Creating regenerative cities. Routledge, Oxon

Lehmann S (2015) Low carbon cities: more than just buildings. In: Lehmann S (ed) Low carbon cities. Routledge, Oxon

Leopold A (1949) A sand county almanac. Oxford University Press, Oxford

Mills G (2015) The urban heat island and low carbon cities. In: Lehmann S (ed) Low carbon cities. Routledge, Oxon

Naess A (1989) Ecology, community, and lifestyle: outline of an ecosophy/Arne Naess. Cambridge University Press, Cambridge

Newman P, Jennings I (2008) Cities as sustainable ecosystems. Island Press, Washington, DC

Olsson K, Haas T (2014) Introduction: emergent urbanism and beyond. In: Haas T, Olsson K (eds) Emergent urbanism. Ashgate, Surrey

Owen D (2009) Green metropolis. Penguin Group, New York

Salingaros NA, Masden KG (2008) Neuroscience, the natural environment, and building design. In: Kellert SR, Heerwagen JH, Mador ML (eds) Biophilic design: the theory, science, and practice of bringing buildings to life. Wiley, New Jersey, pp 59–83

Steiner FR (2011) Design for a vulnerable planet. University of Texas Press

Storey D, Kang KKP (2015) Planning for sustainable and low carbon green cities. In: Lehmann S (ed) Low carbon cities. Routledge, Oxon

Van Den Born RJG, Lenders RHJ, De Groot WT, Huijsman E (2001) The new biophilia: an exploration of visions of nature in western countries. Environ Conserv 28(1):65–75

Wilson EO (1984) Biophilia. Harvard University Press, Massachusetts

Acknowledgements

A visit from Prof. Tim Beatley to my hometown in Fremantle, Australia, ignited the interest and passion to embark on an evolving journey into the emerging concept of biophilic design. Both Tim Beatley and Prof. Peter Newman were instrumental in enabling and supporting the continuing exploration and research travel in North America. Special thanks to both.

Judith Heerwagen from Seattle and Bill Browning from New York, both whom I interviewed on my travels, have continued with their support and generosity of time. Heartfelt thanks to each of you.

Particular gratitude goes to all the people who gave of their time and shared their thoughts in our interviews. They provided the soul of the book. Special thanks go to Scott Edmondson, Tom Liptan, Helena van Vliet, Brad Bass, Cornelia Oberlander, Matt Burlin and Mary Davidge who all helped make the journey such a pleasure by showing me around their cities or helping with accommodation.

Special thanks also to Jett Jackson, my travelling companion, technical assistant and cameraman, who also made it fun.

About This Book

Cities are experiencing rapid population growth and an urban transformation to post-industrial knowledge-based economies. Within this transition is a proposed redefining of urbanites' relationship with nature which suggests that significant human–nature interaction should occur within cities, not just outside. Biophilic design has emerged as a social movement which suggests there is an innate human need to affiliate with nature within the built environment. The book begins by examining the evidence for this innate human need and how this flows into socio-psychological and environmental benefits. These are likely to lead to economic benefits. Together these factors create the basis for a social movement.

Social movements begin with a core group of people uniting with a shared social goal, typically towards change. They are categorised into three stages of development: emergence, coalescence and mainstream. In order to understand biophilic design as a social movement, an immersive journey of heuristic inquiry was undertaken. Heuristic inquiry has four phases which were used to trace the biophilic design social movement:

1. *Focus*—the initial engagement through the literature (Chaps. 1 and 2);
2. *Immersion*—the direct experience in a local and a global journey (Chaps. 3 and 5) involving interviews with key players in the social movement;
3. *Incubation*—the writing of the immersive journal and summarising the interviews (Chap. 3) is combined with reflection on the collected data enabling insight for the final stage; and
4. *Creative Synthesis*—the social movement journey from emergence to coalescence to mainstream is analysed (Chap. 4) in terms of language, the motivators and drivers of change, the barriers and silos preventing change, the interplay between these factors and the steps towards mainstreaming.

The book concludes that biophilic urban design is a social movement moving quickly into the mainstream phase and is completed with a ten-step framework for assisting in the mainstreaming of the social movement of biophilic design based on the conclusions of the heuristic inquiry.

Connecting urbanites to nature through biophilic design appears to encourage a desire for more. The biophilic design social movement journey, if enabled, could ripple through and transform urban landscapes into urban nature habitats.

Contents

About the Author

Dr. Jana Söderlund A childhood spent exploring the beauty of nature fostered a deep passion for nature. Dr Söderlund was a pioneer environmental educator. An early career in environmental science, population and world resources, evolved to further studies in sustainability and climate change; transformative and resilient thinking; and urban design. Since completing her Ph.D. in Biophilic Urban Design, Söderlund has been active as a biophilic designer, presenter, educator and academic author. She is a member of the global Biophilic Cities steering committee, Chair of Biophilic Cities Australia, Director of Biophilic Solutions and Design by Nature, Director of Green Roofs Australasia and a Curtin University associate.

Jana has had an adventurous and diverse life, from science to the arts, and she spent some years backpacking the world, including time living with the Mbuti people in Zaire, Africa and climbing mountains.

A strong passion for nature continues to inspire and motivate her to increase and encourage access for all to the intrinsic benefits of nature in all its patterns, forms, shapes and materials.

Jana Söderlund. Improving mental health in prisons through Biophilic Design. *IFLA World Congress Singapore 2018* pp 34–46

Jana Söderlund, Peter Newman. Improving mental health in prisons through Biophilic Design. *The Prison Journal,* Sage Pub. 2017, http://journals.sagepub.com/eprint/vBW5GG7kibawPAEqA2sD/full

Jana Söderlund. Biophilic urbanism: new opportunities for the horticultural industry. ISHS. Scripta Horticulturae 2016. https://ornamentalhorticulture.emnuvens.com.br/rbho/article/view/1115

Jana Söderlund, Peter Newman. Biophilic architecture: a review of the rationale and outcomes. *AIMS Environmental Science,* 2015, 2(4): 950–969. https://doi.org/10.3934/environsci.2015.4.950

Jana Söderlund, Biophilic Design: a social movement journey, 2016. https://goo.gl/wAvTWP

Jana Söderlund, Biophilic Architecture; Build Australia, 2017. https://www.buildaustralia.com.au/onlineMagazine/APRIL2017/ONLINE.html#p=52

Jana Söderlund, Piloting green walls and studying their effects pp 156–157; Handbook of Biophilic City Planning and Design; Tim Beatley/Island Press, 2016. https://islandpress.org/books/handbook-biophilic-city-planning-design

Dominique Hes, **Jana Söderlund**, Cheryl Desha, Caroline Pidcock. Natural connectors: Biophilic design takes root. *Sanctuary.* Issue 45, pp 69–71.

Cheryl Desha, Anne Roiko, Katharina Nieberler-Walker, Omniya el Baghdadi, **Jana Söderlund**. Building resilience: working across disciplines to foster 'nature and nurture' outcomes. *IFLA World Congress Singapore 2018.* pp 859–869

List of Figures

List of Tables

Chapter 1
The Emergence of a New Social Movement: Biophilic Design

Abstract Biophilic design, based on humans' innate connection to nature, evolved into a social movement over many decades. It emerged from Erich Fromm's early use of the term biophilia to Edward Wilson's book on *Biophilia*, the subsequent *Biophilia Hypothesis* book by Wilson and Stephen Kellert, to the multi-authored *Biophilic Design* publication in 2008. This progression involved meetings and discussions between people with a shared understanding of the human biological connection to nature and the need to bring nature back into urbanites daily lives. Their vision and passion led to the emergence of the social movement of biophilic design. The moral vision of bringing nature into city design reflects a shared understanding and collective goal beyond self-interest. There was a communal sense of the associated issues with cities which lack nature, such as disconnection from nature, urban heat island effect and water and air pollution, kindling a desire to initiate change.

Keywords Biophilia · Biophilic hypothesis · History biophilic design · Social movement · Erich Fromm · Ed Wilson · Stephen Kellert · Biophilic urbanism · Biophilic Cities

1.1 Introduction

This chapter defines the terms biophilia, biophilia hypothesis and biophilic design. Discussion of the origins of the term biophilia as proposed by 1960s psychoanalyst Erich Fromm is important so as to establish the context of where a 'love of life' sits within the human psyche. This understanding and explanation extends the discussion of biophilic design by revealing its deeper dimensions and potential motivators and drivers. The chapter also discusses why understanding social movements is important and how this understanding may contribute towards further strategies of implementation.

By examination of the literature where the terms are introduced, the social movement progression from 'biophilia' to 'the biophilia hypothesis' to 'biophilic design' is revealed. This is through three books. Two of the books resulted from gatherings of interested participants with established background knowledge and contain suggestions of contributing motivators and drivers towards this interest in pursuing and

J. Söderlund, *The Emergence of Biophilic Design*, Cities and Nature,
https://doi.org/10.1007/978-3-030-29813-5_1

developing the topic; the second gathering resulted in the book, *Biophilic Design,* a significant part of the journey as it established the understandings and parameter of biophilic design.

With an understanding of this evolution, it is then possible to establish how biophilic design sits within a social movement framework. The components of a social movement as suggested by social movement theorist Jasper (2007) are outlined and contribute towards the social movement research framework utilised for this book.

1.2 The Origins of Biophilia as a Psychological Term

Biophilia is frequently seen as a term conceived by Nobel Prize-winning biologist E. O. Wilson (Wilson 1984; Kellert and Heerwagen 2008). However, biophilia, meaning love of life, was a term first brought to life by the psychoanalyst Erich Fromm in his exploration of the 'essence of man': that which defines humanity (Fromm 1964). He saw that humans' awareness of their 'beingness', their mortality, separates them from nature, instilling a deep anxiety and conflict. This is a contradiction inherent in human existence, the belonging to two conflicting worlds:

> Man is confronted with the frightening conflict of being a prisoner of nature, yet being free in his thoughts; being part of nature, and yet to be as it were a freak of nature; being neither here nor there. Human self-awareness has made man a stranger in the world, separate, lonely, and frightened.
>
> (Fromm 1964, p. 113)

It is this frightening contradiction and the quest to seek a solution that defines humankind. This search reveals choices, which are reflected in humans' inherent capacity for good or evil.

Early human societal existence tended to regress back to unevolved roots: to animalistic beginnings of nature and violence, to archaic selves, where the fear of the anxiety of separation can disappear. This regressive path is not just in primitive religions but also found in pathological behaviour involving necrophilia and narcissistic and incestuous symbioses. These are tendencies that may be repressed in a collective culture by laws and religions but, interestingly, if the tendencies are supported and shared by many, then consensus gives them credence, reason and reality. This pathway of regression Fromm calls 'the syndrome of decay' (Fromm 1964, p. 114).

The opposite direction, following the 'syndrome of growth', is where the anxiety of separation is faced and transcended, and replaced by a love of life where there is a 'full development of human forces' (Fromm 1964, p. 114). This is the progressive path of life. Fromm suggests that for the majority of people there is mostly a blend of the two orientations.

Fromm saw cities as becoming more mechanistic, industrialised and commoditised, separating humans further from nature and increasing anxiety and even hate for life (1964, p. 10). In his 1964 book, he was expressing concern that, even when there are known extreme consequences, such as in war, there is a stronger societal drive to go to war than to oppose it, which represents the regressive path as the stronger force.

The potential for a triggered eruption of mass violence will always be present as long as there are any human archaic, regressive impulses within. Not until they are fully replaced by a love of life will this potential for regressive behaviour go. At the same time, it is possible for a love of life to be fully replaced by the traits of regression. In either situation, the choice of which path to go down will be lost.

Fromm considers there are three orientations along the progressive path: freedom and independence, love for neighbour and love of life or biophilia (1964, p. 110). Within the biophilic realm is the inherent motivation to live and survive that, as suggested by Fromm and other biologists and philosophers such as Spinoza (Fromm 1964, p. 41), all living substance possesses. Other biophilic qualities include the tendency to integrate and unite, to construct rather than just retain, to adventure to the new rather than stay with the certainty of the past, to use love and reason rather than force and control. "The biophilous conscience is motivated by its attraction to life and joy; the moral effort consists in strengthening the life-loving side in oneself" (Fromm 1964, p. 43).

Fromm outlines the personal conditions he considers necessary for the development of biophilia which include:

- Warm, affectionate contact with others during infancy,
- Freedom and the absence of threats,
- Teaching by example (not preaching) the principles of inner harmony and strength,
- Guidance in the 'art of living',
- Stimulating influence of and response to others and
- A way of life that is generally interesting.

He concludes that being around people who have a love of life will foster a love of life in a child. Fromm identifies the needed societal conditions as security, freedom and justice (Fromm 1964, pp. 47–48). Biophilia, as in a love of life, with a sense of place and connection, provides a solid stepping stone to altruism and love of neighbour. Fromm's recognition that mechanistic and sterile cities contribute to a disconnection from nature leads to an acknowledgement of the potential for cities to become places of nature which nurture the biophilic consciousness.

1.3 Biophilia as a Biological Term

The concept of the biophilic human being was then examined and popularised in 1984 by the sociobiologist Edward Wilson in his book *Biophilia* (Wilson 1984). Here, Wilson defined biophilia as "the innate tendency to focus on life and lifelike processes" (Wilson 1984, p. 1). He utilised the term biophilia to describe his deep feelings of connection to nature during a period of exploration and immersion in the natural world. Wilson's exploration of the term biophilia deepened Fromm's writings; these two thinkers agreed that a biophilic love of life is inherent in human beings, is able to be nourished by connecting with nature and that, if the opportunity to foster a biophilic love of life is not present in a human's life, then the tendency is to follow a regressive, more destructive life path. Wilson's unique proposal was that this biophilic propensity developed as part of evolutionary survival and thus encompasses certain characteristics.

Wilson observed that there were common landscape features that attracted him, raising questions about whether there were preferred habitats for humans, as displayed by animals. He hypothesised that, if given a choice, there are key features that humans would select, features similar to the original evolutionary environment of savannah grasslands and supplementary features that aided survival. These he listed as the savannah itself, vantage points and water bodies (Wilson 1984, p. 110).

When given free choice, Wilson observes, people build on vegetated, higher ground near water but, when unable to, attempt to recreate a similar terrain, an action he labels 'savannah gestalt' (Wilson 1984, p. 111). In crowded cities, small gardens with fountains and pools flourish alongside trees and shrubs, often pruned to heights and shapes similar to the tropical savannah.

Through his writings Wilson was able to provide an evolutionary biologist's view of biophilia as something that humans have evolved to need, arguing that human existence, and mental development, depends on this propensity. He concludes his book suggesting:

> …as biological knowledge grows the ethic will shift fundamentally so that everywhere, for reasons that have to do with the very fiber of the brain, the fauna and flora of a country will be thought part of the national heritage as important as its art, its language, and that astonishing blend of achievement and farce that has always defined our species.
>
> (Wilson 1984, p. 145)

1.4 Developing the Biophilia Hypothesis

Wilson received significant response to his book. His narration of a personal journey brought the term biophilia to life, with a biological focus that was easier for many to relate to than Fromm's psychological perspective. As Fromm first postulated, affiliation with nature can assist in humans' intellectual, emotional and spiritual fulfilment; Wilson then expanded this to suggest that evolution has impelled humans' biological design to respond positively to this contact with nature (Wilson 1993).

Scholars from diverse fields collected together nearly a decade later, in 1992, to collaborate their thinking and debate the concepts presented by Wilson. From this assemblage of intellectuals emerged the book *The Biophilia Hypothesis*. In Stephen Kellert's introduction to the book, he suggests that a much richer and deeper book resulted from "the highly attractive, retreat-like setting and the highly productive conversation" (Kellert 1993, p. 24). The challenge with the book was to sufficiently examine evidence to build a foundation to support future inquiry into the suggested assertions of the nature of humans' desire to connect with life and lifelike processes. The biophilia hypothesis proposes that the qualities of this desire are that it is:

- Inherent (that is, biologically based)
- Part of our species' evolutionary heritage
- Associated with human competitive advantage and genetic fitness
- Likely to increase the possibility for achieving individual meaning and personal fulfilment
- The self-interested basis for a human ethic of care and conservation of nature, most especially the diversity of life.

(Kellert 1993, p. 21)

1.4.1 The Collective Focus

The group who came together to discuss and ultimately write the *The Biophilia Hypothesis* were invited by Wilson and Kellert on the basis of the relevance of their prior work (Kellert 1993). Kellert expressed in his introduction the importance of not having the book dismissed as promoting an idealisation of nature (p. 21), but seeing it as a scientific-based inquiry. He also recognised that they were building on previous knowledge in the area, though not labelled as biophilia, and that the "richness of the topic requires no less than a multidisciplinary consideration" (p. 22). The book indeed reveals a diverse collection of authors who all are, in varying ways, connected with nature. Nature was the thread that brought them together, but, in all of their writings, they have recognised some form of disconnect from nature that pervades current urban society, and of something lost through this separation. This united discontent with modern cities supported the acknowledgement of the need for change in contemporary lifestyles. According to social movement theorists (Jasper 2007; Christiansen 2009), the first stage of formation of a social movement is with a collective discontent and desire for change.

Particular author chapters reflect this more than others; Orr titled his chapter on biophobia and the obstacles to biophilia, '*Love it or lose it: The coming biophilia revolution*' (Orr 1993), but the concluding chapter by Soulè, '*Biophilia: The unanswered questions*', a commentary on the discussions that took place in the prelude to the book's publication, is particularly revealing of the group's sentiment and desire for change (Soulè 1993). Soulè suggests that, like notions of non-violence or civil disobedience, the idea of biophilia could bring about change, but he thinks that social inertia will retard the move, as has happened historically in other social movements (Soulè 1993). He suggests that:

> If biophilia is destined to become a powerful force for conservation, then it must become a religion-like movement. Only a new religion of nature, similar but even more powerful than the animal rights movement, can create the political momentum required to overcome the greed that gives rise to discord and strife and the anthropocentrism that underlies the intentional abuse of nature.
>
> (Soulè 1993, p. 454)

1.5 The Emergence of Biophilic Design

Even though there was a strong, positive reception of the book *The Biophilia Hypothesis,* it was 14 years later, in 2006, that the next progression of the biophilia hypothesis truly occurred, through another gathering of people. It took place in a beautiful setting on Rhode Island when Kellert brought together a diverse group including academics, industry representatives and real estate investors who had shown a shared interest in increasing the opportunity for nature to find expression in urban design. Many ideas were discussed and design principles emerged. These resulted in a book called *Biophilic Design: The Theory, Science, and Practice of Bringing Buildings to Life* (Kellert et al. 2008), which introduced the concept and rationale for nature no longer being minimised in cities but recognised as having much to offer in everyday urban life and therefore design. The book presented the birth of biophilic design and articulated a new approach in thinking and design by defining an alternative human–nature connection for urban inhabitants. Biophilic design was summarised as "the expression of the inherent human need to affiliate with nature in the design of the built environment" (Kellert and Heerwagen 2008, p. viii).

In the preface, Kellert and Heerwagen define biophilic design as:

> The expression of the inherent human need to affiliate with nature in the design of the built environment.
>
> (Keller and Heerwagen 2008, p. viii)

Kellert and Heerwagen saw that contributors to the book offered more than strategies for sustainability but also strategies for achieving "a more satisfying and fulfilling modern society in harmony with nature" (Keller and Heerwagen 2008, p. vii). This, they then suggested, would be a paradigm shift in how we design and build. Berkebile et al. (2008) encapsulate this. They suggest that biophilic design provides "a new language for interpreting the built environment" (p. 347) and "biophilia represents an abundantly creative moment in design" (p. 349). They discuss the possibility for a new ethic for excellence in design with the hope that in the near future green building design will no longer be a specialty field but that the biophilic design movement has "driven a fundamental shift" (Berkebile et al. 2008, p. 349), perhaps one that would also offer greater potential to follow the biophilic progressive path of life.

The social movement of biophilic design was emerging.

1.5.1 Biophilic Urbanism

Within the book, *Biophilic Design*, was a contribution by Tim Beatley from the University of Virginia which integrated the idea of biophilia at the urban scale as biophilic urbanism. This differs from just designing nature into the city by including consideration of the human–nature connection and the need for humans to have a daily interaction with nature. Beatley began to draw together ideas from around the world and from other academic literature to show that the concept needs to be mainstreamed in cities. His book *Biophilic Cities* in 2010 and the subsequent book *Handbook of Biophilic City Planning and Design* in 2017 are accompanied now by a website that services a Biophilic Cities Network. This network has been created to support a growing social movement of cities around the world that are making biophilic design part of their policy framework and practice.

This social movement is paving the way for a redefining of urbanites' relationship with nature, utilising the term biophilia introduced by Fromm in the 1960s. To further understand this new philosophy, or cultural shift, we need to understand the features of such a social movement.

1.6 Social Movement Theory

Social movements have been studied since the French revolution and era of intellectuals and philosophers such as Kant, Rousseau and Hegel, but Lorenz Von Stein is given credit for introducing the term (Brown 1966; Cahnman 1966). Understanding of social movements is significant as they are the essence of social life: "They provide a framework for people coming together to solve problems. Time, energy and money will be devoted to a collective purpose, actions will be coordinated. This is what social life is all about. Movements help us ask, what are people like, and why do they do what they do? How do they get things done? How do they create something new?" (Jasper, personal communication, 2015).

As global cultures evolve and change through history so do societal structures and social needs and causes. Throughout history people have united through discontent with an aspect of their society, with the intention of progressing a change. Like thinkers come together to design and implement strategies, they create a social movement for change. The manner in which the unification, the strategies and the implementation of change occurs has been flexible and responsive to the societal conditions and the causes being championed. Definitions of social movements have tended to be as elastic as the structures. The Online Encyclopaedia Britannica provides a contemporary general and workable definition:

> **Social movement**, loosely organized but sustained campaign in support of a social goal, typically either the implementation or the prevention of a change in society's structure or values. Although social movements differ in size, they are all essentially collective. That is, they result from the more or less spontaneous coming together of people whose relationships are not defined by rules and procedures but who merely share a common outlook on society.
>
> (Smelser n.d.)

1.6.1 The Changing Nature of Movements

As recently as the 1950s, social movement groups were typically seen as involving violent, active protest by angry, oppressed peoples, often psychologically damaged and disconnected. Since the mid-1960s, after the success of the black civil rights movement, the 'psychological' model of the 1950s was replaced with an economic model, comprising movements such as the animal rights movement, environmental movement or the women's movement. These movements tended to be portrayed as more sympathetic movements, focused on their own beliefs, and were never quite given respect for their contribution to the greater good of society (Jasper, 2015, personal communication).

Today's communication technologies, particularly the Internet, permit social connection on an unprecedented scale, which has allowed social movements to form in new ways. The vast diversity of global interests is well presented, and easily found. It is rare to search for a topic and not find some reference to the subject. It has become easy to discover other people with similar perspectives and attitudes on life. This has enabled social networks to evolve globally at a rapid pace. Online social movements such as Avaaz and 350.org have become large and powerful voices. They both began with a cause and a group of passionate people. Avaaz define themselves as "a global web movement to bring people-powered politics to decision making everywhere" (http://www.avaaz.org/en).

These social movements are structured and function in vastly different ways to historical social movement groups which tended to be confined within a cultural group. Social networks are now easily formed on a global scale.

1.6.2 The Social Movement Theory of James Jasper

The changing nature of social movements in a digital age of social media has led to new and varying images and theories of social movements (Sheffield 2010). The social movement theory of Jasper (2007) has been adopted as the framework for this study as it provided a basis for how rapid transition from idea to movement can happen.

He outlines three categories of components of contemporary social movements: micro-foundations, strategies and emotions, plus four features which delineate social movements: intention, coordinated action, persuasive tactics and that they are sustained over time.

Social movements can also be viewed as progressing through stages (Jasper 2007; Christiansen 2009). I suggest that social movements are generated internally and are then increasingly externalised in three stages:

- Emergence—where people with similar moral visions and emotions come together to plan strategies for change. This stage involves the internal motivators.

- Coalescence—clear strategies have emerged, leaders have emerged and the movement is gaining momentum. This stage reveals the drivers to start change.
- Mainstream—the philosophy of the movement is adopted and appearing in everyday literature, policy and other mainstream initiatives. In this stage, the policies for implementation are formed.

Utilising the identified components of social movement theory as outlined by Jasper (1998, 2007) and Christiansen (2009) provides a framework for analysis of the social movement of biophilic design. This framework was utilised throughout the immersive journey in interviews with key players in Australia and North America.

Where encountered, the following components were identified:

Players—the interviewees.
Arenas—where the players are situated: academia, government, industry (biophilic design industry and profession) or civic.
Strategies—what they are using to implement biophilic design: policy, research, application, innovation, etc.
Motivators and drivers—what urges the players to be involved in biophilic design initiatives.
Stages—emergent, coalescent or mainstream.
Barriers—any difficulties which players have encountered in their journeys.

1.7 The Social Movement of Biophilic Design

The emergence of biophilic design has involved a collection of people who are deeply embedded and active in their culture, jointly recognising that there is a need to change the approach to designing cities. The moral vision of bringing nature into city design reflects a shared understanding and collective goal beyond self-interest.

Since the initial 2007 conference where biophilic design as a term was established, strategies have intentionally and non-intentionally been utilised to promote it. Published research in the form of papers and books have progressed and supported the movement. These have either focused specifically on biophilic design, or on research that is in support of the concept and the need for a new approach to urban lifestyles. Websites, blogs and conferences have disseminated information on biophilic design. Implementation of policy has been a strategy successfully employed in some cities, particularly in the United States, as a tool to coerce people to build with biophilic initiatives. Another strategy has been to refine and promote ways of implementing biophilic design, the how-to guides and ideas, with both current and historical examples.

Emotions need to be examined as part of a movement's own dynamics (Jasper 1998). They are tied to moral values, beliefs and assumptions and provide the motivation for responsive action. Emotions are processes that not only involve thinking and feeling but all things happening in human bodies. Emotions can give meaning to the "something that resonates" (Jasper, personal communication, 2015). Emotions as a component of the social movement of biophilic design may be significant where they are identified. Biophilia's root origins were with the emotion of love, the term biophilia meaning 'love of life'.

Collecting individuals' stories is an important contribution to revealing emotions and motivators, but to gain a wholistic perspective of the social movement the individual responses need to be aggregated. This helps to identify the collective identity of the social movement. Even with individual interests a sense of solidarity can exist between members with a clear collective identity. This involves "aligning of frames" (Jasper 1998). Within social movements cognitive agreement alone does not lead to action; aligning of motivational frames is also important, giving the motivation and emotion to protest. There may be a variety of motivators, dependent on the player and the arena, yet with a deeper commonality that encourages people to get involved. The common problem must be identified and the common prescription for solving it. Collective rites are frequently part of the collective identity and they can help clarify frame alignment. These bring together the main players/characters but also allow new players to connect with the movement and join the collective identity. Part of the collective identity is the recognition not only of the benefits but also of the risks, which present as barriers to the implementation of biophilic design.

Proponents of biophilic design may feel strongly about the current way humans design and inhabit cities. There may be a communal sense of injustice towards biophobic cities which deny their inhabitants meaningful contact with nature. As Jasper suggests, social movements need the victims and the villains and the heroes (Jasper, personal communication, 2015). In this framework, biophobic cities and the associated issues such as disconnection from nature, urban heat island effect and water and air pollution are the villains. Distressed inhabitants and the natural environment are the victims. The heroes of biophilic design may be the local champions who are striving to implement design elements within their arena.

1.8 Conclusion

Fromm utilised the term biophilia to describe a love of life that is beneficial in the human psyche in the quest for 'wholeness' and spiritual fulfilment. By reviewing the early literature on biophilia and biophilic design, this chapter confirmed biophilic design as new social movement driven by a collection of people with the common aim of reconnecting people to nature in an urban environment, with all the social benefits that accompany this. They were motivated by discontent with the current disconnect from nature that can occur within cities and they wanted to change this. The groups which gathered shared experiences and creativity alongside a desire to create healthier and more pleasing cities.

The book, *Biophilic Design: The Theory, Science, and Practice of Bringing Buildings to Life* (Kellert et al. 2008), developed the ideas in *The Biophilia Hypothesis* (Kellert 1993). The theory, the science and the design elements presented in the book focus on the human–nature connection and the social benefits this brings. Social movement theory has provided a rationale for understanding biophilic design as a social movement.

The following chapter will examine the research literature which supports the theory and assertions of biophilic design and the benefits identified as a worthy design approach for cities.

References

Berkebile B, Fox B, Hartley A (2008) Reflections on implementing biophilic design. In: Kellert SR, Heerwagen J, Mador M (eds) Biophilic design: the theory, science, and practice of bringing buildings to life. Wiley, Hoboken, N.J

Brown B (1966) The French Revolution and the rise of social theory. Sci Soc 30(4):385

Cahnman WJ (1966) Lorenzo Von Stein: the history of the social movement in France, 1789–1850. (Book Review) Am J Sociol 71:746–747

Christiansen J (2009) Four stages of social movements. Research Starters: Academic Topic Overviews, EBSCO, Ipswich, MA

Fromm E (1964) The heart of man. Harper and Row, New York, USA

Jasper J (1998) The emotions of protest: affective and reactive emotions in and around social movements. Sociol Forum 13(3):397–424

Jasper J (2007) Social movements. In: Ritzer G (ed) The Blackwell encyclopedia of sociology. Blackwell, Malden, MA

Kellert SR (1993) Introduction. In: Kellert SR, Wilson EO (eds) The biophilia hypothesis. Island Press, Washington, DC, USA

Kellert SR, Heerwagen J (2008) Preface. In: Kellert SR, Heerwagen J, Mador ML (eds) Biophilic design: the theory, science, and practice of bringing buildings to life. Wiley, Hoboken, N.J

Kellert SR, Wilson EO (eds) (1993) The biophilia hypothesis. Island Press, Washington, DC

Kellert SR, Heerwagen J, Mador ML (eds) (2008) Biophilic design: the theory, science, and practice of bringing buildings to life. Wiley, Hoboken, NJ

Orr D (1993) Love it or lose it: the coming biophilia revolution. In: Kellert SR, Wilson EO (eds) The biophilia hypothesis. Island Press, Washington, DC, USA, pp 415–440

Sheffield J (2010) Weblogs and activism: a social movement perspective on the blogosphere. In Hogan JM (ed). ProQuest Dissertations Publishing

Smelser NJ (n.d.) Social movement. http://www.britannica.com/EBchecked/topic/551335/social-movement. Accessed 09 Sept 2014

Soule ME (1993) Biophilia: unanswered questions. In: Kellert SR, Wilson EO (eds) The biophilia hypothesis. Island Press, Washington, DC

Wilson EO (1984) Biophilia. Harvard University Press, Massachusetts

Wilson EO (1993) Biophilia and the conservation ethic. In: Kellert SR, Wilson EO (eds) The biophilia hypothesis. Island Press, Washington, DC

Chapter 2
The Rationale for Biophilic Design

Abstract Biophilic design helps create regenerative, resilient and happy cities. Contemporary cities are under stress. They have been built as human habitats, but designed by economics, not by liveability, well-being, or as a healthy, sustainable habitat for humans. Current pressures of increasing urban density and climate change are amplifying crises of human health such as stress, obesity and mental health issues, along with the heat island effect and water and air pollution. Biophilic design, based on humans' innate connection to nature, is a rapidly expanding urban design approach initially driven as a response to these urban crises, but now being taken up by progressive and innovative cities who acknowledge the multiple social, environmental and economic benefits which result. Consistently emerging research of the responses to the rich and diverse biophilic elements (natural forms, patterns, places, materials, greenery) shows the endless remedial possibilities of inserting cost-effective biophilic elements and uniting nature and the urban fabric. This brings both physical integration and human connection and values, solving the problem of habitat fragmentation and human disconnect resulting from underutilised spaces and incompatible built elements.

Keywords Environmental psychology · Biophilic benefits · Biophilic design · Biophilic attributes · Stress reduction · Fractal patterning · Attention restoration

2.1 Introduction

The previous chapter introduced biophilic design as a social movement driven by scholars and others who united in advocating for the benefits that this design approach could bring to cities. This chapter continues the story of the emergence of the biophilic design social movement through examination of the research and insights that were being revealed through better technologies and research methods. This research was a crucial component to the emergence of biophilic design as it provided the validity for the approach. The chapter outlines some of the problems the lack of nature in contemporary urban design approaches bring. The early research which examines

J. Söderlund, *The Emergence of Biophilic Design*, Cities and Nature,
https://doi.org/10.1007/978-3-030-29813-5_2

the innate biological human responses to nature from both a psychological and physiological perspective is reviewed. It was these timely insights which contributed to helping define the emerging design principles and attributes of biophilic design.

These biophilic design principles, which are based on the human evolutionary connection and support the greater inclusion of nature in cities, are outlined. Where implemented, social and environmental benefits result. These combine to provide potential economic benefits, though there is little direct research on this. The need to understand the potential multiple benefits that biophilic design initiatives can yield is highlighted.

To gain a clearer perspective on what design elements and what benefits are receiving most attention, the chapter concludes with an analysis of the literature.

2.2 The Current Urbanity

A core theme from the biophilic design literature is that humans have lost something in their approach to building design in modern times (Heerwagen and Gregory 2008). Human affiliation with nature is seen to be historically reflected in organic building designs and materials, in patterning and spaces that mimic those of nature and in traditional living in close but respectful proximity to the natural environment. Greening of roofs and walls was commonplace in traditional architecture, providing insulation, food and aesthetics. However, modern architecture has lost this scope. Advancements in industry and technology not only provided the means to mechanise and sterilise buildings and design, but also the ability to influence humans' psychological attitude to nature. Salingaros and Masden (2008) proposed the three "conceptions of human beings":

1. The Abstract Human Being – humans are regarded as a component in a mechanical world.
2. The Biological Entity – the human being is an organism made of sensors that interact with its environment.
3. The Spiritual Being – humans are something more than a biological neural system, connected to the universe in ways that other animals are not.

(Salingaros and Masden 2008, pp. 66–67)

They suggested the contemporary, mechanistic and sterile urban world has fostered an increase in the abstract human being, of people living without direct connection to nature. They live as "an inert passenger in a fundamentally sterile and non-interactive world" (Salingaros and Masden 2008).

Societies have been able to surpass the need to accommodate direct connections with nature, modifying the environment to perceptually suit human needs while encouraging a separation and disconnect from the natural world (Miller 2005; Orr 1993). The intuitive more humanistic designs of human history were left behind. Modernist designs encouraged a fear of nature (due mostly to waterborne diseases and parasites) that led to very sterile urban environments (Newman and Matan 2012).

There is significant academic and practice-based literature on the health, including mental health, of contemporary cities (e.g. the WHO Healthy cities which began in 1986). The contention is therefore made by biophilic design writers that contemporary cities are places not designed for mental health and well-being (Hes and Du Plessis 2015). Biophilic designers see the missing evolutionary element in modern cities as a need to re-establish an innate connection to nature in everyday life. As Tim Beatley says "we need a daily dose of nature", which means nature must be integrated into all parts of buildings, not separating people in buildings from people in nature (Beatley 2011). They suggest that as biological beings, humans have not adapted physiologically, emotionally or psychologically to the current sterile urban technological cities. This 'mis-match', where the environment humans live in is so alien from the one from which they have evolved, is thus hypothesised to lead to much of modern stress and to mental health issues (Nesse and Williams 1995; Burns 2005).

2.3 An Evolution of Thinkers and Ideas

The ability of architectural design to influence individuals' physiological and psychological states is an extension of the biophilic connection to nature. Expression of this connection through biophilic design in architecture has occurred throughout history, not always consciously or even acknowledged, conveying a subjectiveness which testifies to its inherent quality in humans. Nature can be mimicked by using patterning, forms, materials, symbols and spaces which represent nature and evoke similar responses. Christopher Alexander recognised this in his seminal book Pattern Language; though not using the term biophilia, he expressed similar insights:

> Many of the patterns here are archetypal – so deep, so deeply rooted in the nature of things, that it seems likely that they will be a part of human nature, and human action, as much in 500 years as they are today.
>
> (Alexander et al. 1977)

As with biophilic design theory, Alexander believed that the pattern language of the nature of things in the environment "can make people feel alive and human" (Alexander et al. 1977). When people cannot surround themselves with nature, then architecture and landscapes that contain some archetypal natural elements have found expression in urban design.

Similar to Wilson (1984), Appleton (1975), a landscape architect, considered that humans' aesthetic reactions to landscape and architecture "are in part inborn" and therefore people cannot stray too far from the natural patterning before destroying their "aesthetic experience" (Appleton 1975, p. viii). He posited that humans must seek to recreate something of the 'primitive connection' with nature to maintain an experience of well-being. The prospect-refuge theory conceived by Appleton suggests that individuals feel good when safe in a place of refuge, a feeling enhanced when they have a window overlooking life and the happenings around them. This reflects the innate protective need to survey for hazards from a place of safety (Appleton 1975). Either refuge or prospect on its own can still contribute to a sense of well-being but Appleton suggests the two together are the most appealing.

Wilson (1984) had also considered prospect (vantage points) and refuge attributes that contribute to positive human feelings. In addition, he theorises that humans' evolutionary beginnings in the African savannah similarly led to a positive psychological response to environments with shade trees, waving grasses and far vistas. This is now commonly known as the 'savannah effect' and is demonstrated in a design example by Frank Lloyd Wright with his iconic Johnson Wax building.

2.4 Evidence from Human Psychology and Physiology

In the last 30–40 years, technological advancements in both psychological and physiological testing have enabled further exploration and testing on whether there is an innate human relationship with nature which is the fundamental rationale for biophilic urbanism. This research is steadily, and rapidly, increasing, as scientific evidence continues to compound through new techniques and studies which support the intuitive knowing of early and historic advocates.

In 1979, Ulrich began investigating links between psychological well-being and physiological responses when individuals are exposed to nature, or even views of nature. Psychological testing of responses to projected slides revealed that stressed individuals feel considerably better when exposed to views of nature (Ulrich 1979). In 1984, Ulrich decided to test this response with hospital patients' analgesic usage and recovery times with and without a view of nature. Recovery times were faster for the patients with the view to nature, with less need for pain relief (Ulrich 1984).

Kaplan and Kaplan were also researching the potential benefits of the human relationship with nature and in 1989 presented a psychological perspective of experiencing nature. They built on the 1892 work of William James who had identified two types of attention, voluntary and involuntary. Involuntary attention is what is given to things that catch the eye, often moving, patterned, bright and stimulating. Voluntary attention is where we hold a focus, blocking out unwanted stimuli, and it can be exhausting. Aggressive, irritable and antisocial behaviour plus slow responses can result from directed attention fatigue. Restoration is important and involves involuntary attention. Kaplan and Kaplan suggested that exposure to nature over a range of environmental choices can fulfil the criteria for directed attention or voluntary attention restoration (Kaplan and Kaplan 1989).

In 1991, Ulrich et al. employed electrocardiograms (EKG) and measured pulse rates, frontalis muscle tension and skin conductance plus self-ratings of emotional states to further investigate the physiological relationship with nature. Both physiological and verbal results indicated that recovery from stress was faster in a natural setting than an urban one. The physiological results also suggested an involvement of the parasympathetic nervous system (Ulrich et al. 1991). Ulrich et al. proposed a psycho-evolutionary theory that nature restores through increasing positive feelings, positive physiological responses and sustained involuntary attention. Ulrich's stress reduction theory was perceived to contradict the Kaplans' restorative theory which suggests that peoples' directed attention relaxes in nature due to an involuntary (fascination) attention and thus is restorative. Kaplan's investigation of this apparent dichotomy led to an integrative understanding that deepened the theoretical exploration of the human–nature connection and formed the Attention Restoration Theory (Kaplan 1995).

The biochemical underpinnings of human psychological and physiological responses were a little examined area until the work of neuroscientist Candace Pert and her colleagues. Pert's discovery in 1972 of humans' opioid receptors (Pert and Snyder 1973) and the subsequent discovery of the natural opiate of enkephalin (endorphin) by Hughes and Kosterlitz (Pert 1997, p. 64) pioneered the ability to test for feelings of pleasure and well-being. The link between psychological well-being and physiological responses became established.

Since Pert's early work other physiological markers of psychological feelings and moods have been established, enabling quantifiable physiological analysis. Cortisol and cortisone are hormones released when the body is stressed. Cortisol is now widely used as a stress marker. Blood pressure, heart rate, skin moisture conductivity all increase when individuals are anxious or stressed. Studies have emerged from Japan on the effects of the traditional Shinrin-yoku, or forest bathing. These and other research have shown that exposure to nature reduces heart rate variability and pulse rates, decreases blood pressure, lowers cortisol and increases parasympathetic nervous system activity while decreasing sympathetic nervous system activity (Berman et al. 2008, 2012; Li et al. 2011; Matsunaga et al. 2011; Park et al. 2010; Tyrväinen et al. 2014). These responses contribute to improved cognitive functioning, working memory and learning rates. Forest walking has also revealed that levels of the hormone DHEA tend to increase (Li et al. 2011).

Both Kaplans' and Ulrich's theories have been put to the test in the years since they were first proposed, either directly or by studies revealing supporting results. Berto (2005) undertook three experiments involving 32 participants, concluding that restorative environments and experiences that involve nature do greatly support mental fatigue recovery. She suggests that in a 'world overflowing with information', mental fatigue is endemic and much could be done, especially in institutions, to help cognitive well-being (Berto 2005). Following the increasing interest in Kaplan's restoration theory, Tenngart Invarsson and Hagerhall (2008) also began investigations into the restoration values between differing forms of natural environments among built environments, such as gardens. The varying results between gardens suggested that greater understanding of the form of the natural environment, and its potential to be restorative, was needed (Tenngart Ivarsson and Hagerhall 2008).

Salingaros and Masden suggest that "environments devoid of neurologically nourishing information mimic signs of human pathology. Drab minimalist surfaces reproduce symptoms of strokes and macular degeneration, for example" (Salingaros and Masden 2008, p. 69). Environments that are devoid of any representation of nature can not only make people psychologically unwell and regressive in their behaviour but people can also display physical symptoms and responses. A recent study which examined human responses to design stimuli concluded that the primal flight or fight response is increased when individuals are exposed to hard-edged architecture rather than curving contours (Nanda et al. 2013). They also suggested that this response is heightened when a person is already in a stressful environment such as a hospital (Nanda et al. 2013). Hartig et al. (2008) support the theoretical evidence for restorative environments. They discuss the relationship between restorative environmental design and biophilic design suggesting that restorative design encompasses more than biophilic design by taking into account low-impact technologies, people's activity cycles and varying needs for restoration plus the impact of cultural experience on people's receptivity to biophilic influences.

Increasing greenery in housing estates resulted in less violence and aggression, less crime and better interpersonal relationships (Kuo and Sullivan 2001). Further research by Kuo also suggested that greener environments in poorer public housing estates reduce mental fatigue and assist "residents' psychological resources for coping with poverty" (Kuo 2001). Studies by Guègan and Stefan observed that short immersions in nature elicited a more positive mood and a greater desire to help others (Guéguen and Stefan 2014).

Berman et al. (2008) investigated the interaction with nature on direct attention restoration and improved cognitive functioning by comparing urban and natural environments. Results further validated Kaplan's restoration theory and showed that even viewing pictures of nature can improve cognitive functioning, mood and working memory (Berman et al. 2008). Raanaas, Evensen, Rich, Sjøstrøm and Patil conducted controlled laboratory experiments on attention restoration with and without plants. Though only four pot plants were utilised, there were improvements in performance in the room with the plants compared to the room without (Raanaas et al. 2011). Their results suggest that exposure to nature could be a valid supplement to treating depression and other disorders, with improvements to mood and memory span (Berman et al. 2012; Tyrväinen et al. 2014).

Park and Mattson (2008) suggested that, with further supporting evidence, plants should be used in hospitals as a supplementary healing mode. Their research had confirmed Ulrich's early studies on the positive effects that nature has on a patient's recovery period and analgesic need. Park and Mattson found that indoor plants "enhance patients' physiological responses, with lower ratings of pain, anxiety and fatigue, and more positive feelings and higher satisfaction with their hospital rooms" (Park and Mattson 2008). A study in Michigan revealed a 24% less frequency of health care visits for prison residents with views of nature (Moore 1981). Measurements of elderly women exposed to a green rooftop forest on a hospital showed that they were more physiologically relaxed and restored (Matsunaga et al. 2011).

Research on the physiological and psychological responses of office workers to a vase of roses by Ikei et al. (2014) demonstrates the use of both psychological and physiological markers. Heart rate variability, pulse rate and subjective responses, evaluated through a Profile of Moods (POMS) questionnaire, were measured. Heart rate variability is a physiological indicator of the human nervous system and thus can be used to reflect parasympathetic and sympathetic nervous system activity (Li et al.

2011; Matsunaga et al. 2011; Park et al. 2010). The study showed that by simply viewing roses, parasympathetic nervous system activity increases, indicating lower stress and a greater sense of well-being (Ikei et al. 2014).

A Finnish study investigated the psychological effects (restorativeness, vitality, mood and creativity) and the physiological effects of short-term immersion in nature (Tyrväinen et al. 2014). The physiological response was measured using salivary cortisol as an indicator of stress. Results suggested that even short-term exposure to nature had positive effects on stress compared to the urban built environment (Tyrväinen et al. 2014).

Nieuwenhuis et al. (2014), noticing two opposing trends in offices, conducted studies of the comparison: lean versus green. They concluded that lean is "meaner than green", not only because it was less pleasing to the workers but also because organisational output and productivity was significantly less in the lean offices.

It is not just direct exposure to greenery that has positive human responses. Research by Tenngart Ivarsson and Hagerhall (2008) suggested that there may be different human responses to different natural forms. People respond both psychologically and physiologically to natural patterning and the spaces of nature. Prime among these are the fractal patterns of nature (self-replicating patterns that occur at increasingly smaller magnification and are found throughout nature), especially those with "high randomness and mid to low fractal dimension". Research suggests that these patterns relax and de-stress people (Hagerhall et al. 2012).

Viewing nature, especially the richer patterns, is literally pleasurable due to the stimulation of the mu-opioid receptors in the human brain and greater endomorphin release (Biederman and Vessel 2006). Although seen throughout architecture and art and intuitively appreciated, it is only recently that measurement of the psychological and physiological responses to fractal patterns has occurred, enabled by Pert's earlier and significant discovery (Ode et al. 2010; Stamps 2002; Taylor 2006). It was found that certain fractal dimensions trigger more intense physiological responses, with many of these responses indicative of stress reduction (Taylor 2006). Taylor suggests how incorporating a rich variety of fractal patterns into buildings can be useful in situations where "people are deprived of nature's fractals" (2006). Varying sounds, colours and light can produce similar pleasurable physiological responses, as can movement such as waving grasses, especially when viewed in the eyes' periphery.

Thus, there is strong evidence for an innate human response to nature. Psychological and physiological evidence is continually emerging confirming the scientific base to biophilic design and the assertion that humans respond positively to nature and the patterns and spaces of nature.

2.5 The Environmental Benefits

Decreased biodiversity, urban heat island effects and pollution have become current and urgent environmental issues that challenge the resilience of cities. Kellert, Heerwagen and others in the biophilic design movement from the 2006 conference and beyond recognised the potential environmental benefits of restoring and enhancing nature in urban design. These emergent biophilic design advocates primarily focused on the human–nature connection though they acknowledged the possible benefits, not only with the human–nature relationship but also environmentally that a shift towards a design approach that integrates nature into cities could bring. Following this time, the biophilic design movement, especially Beatley (2011) and his research group (http://biophiliccities.org), has emphasised environmental restoration and regeneration.

In the last decade, research on the environmental benefits of biophilic design has focused on the benefits of direct greenery, predominantly green roofs, living walls and rain gardens. Green roofs have a historical place in urban design with the early sod roofs of European architecture, but have developed new engineering techniques to enable green roofs to become a major architectural feature of innovative buildings (Tan 2013). Vertical greenery has also progressed from vine-covered facades to vertical living walls since the aesthetic designs and constructions of innovative French botanist Patrick Blanc (Beatley 2011; Yok et al. 2009). As a result, a range of environmental benefits have been evaluated including improvements to water, air, biodiversity and heat. These are detailed in a paper by Söderlund and Newman (2015) which reviews biophilic design attributes and the resulting benefits.

2.6 Biophilic Design Principles

The historic design principles and understandings of people such as Alexander and Appleton, coupled with the continually emerging supporting science and new green infrastructure techniques, provided a solid design basis for including nature in the built environment.

Proponents of biophilic design elaborated on these design concepts, finding validity through experience, intuitive knowing and historical examples (Heerwagen and Gregory 2008; Kellert 2008a, b; Wilson 2008). Authors in the book, *Biophilic Design: The Theory, Science, and Practice of Bringing Buildings to Life* (Kellert et al. 2008), recognised the need to define the dimensions of biophilic design as the beginning of a tool kit for urban designers and developers. Heerwagen and Gregory (2008) categorised seven major attributes whereas Kellert (2008a, b) listed six elements with 70 design attributes.

Ryan et al. (2014) refined these design elements of biophilic design with supportive qualitative and quantitative research in both the physiological and the psychological. Browning, recognising previous design attribute lists were unwieldy and potentially confronting for urban designers, consolidated the design attributes to the following 14 patterns within three categories (Table 2.1).

Kellert later revised and simplified his 70 design attributes. Twenty-four design attributes were headed by three categories of experience similar to Ryan et al.: direct experience of nature, indirect experience of nature and experience of space and place (Kellert and Calabrese 2015).

Table 2.1 Patterns of biophilic design

Nature in the space incorporation of plants, water and animals into the built environment, especially with movement	***Natural analogues*** one degree of separation away from true nature; patterns and materials that evoke nature	***Nature of the space*** the way humans respond psychologically and physiologically to different spatial configurations
1. Visual connection with nature—plants inside and out, green roofs and living walls, water, nature artwork	**8. Biomorphic forms and patterns**—organic building forms, structural systems (savannah effect)	**11. Prospect**—views, balconies, 6 m and above focal lengths, open floor plans
2. Non-visual connection with nature—sun patches, textured materials, bird sounds, weather, nature scents	**9. Material connection with nature**—wood, earth and stone construction, natural colours	**12. Refuge**—protected spaces, overhead canopies or lowered ceilings, places providing concealment
3. Non-rhythmic sensory stimuli—clouds, shadows, nature sounds, water reflections	**10. Complexity and order**—fractal patterns, sky lines, plant selection and variety, material textures and colours	**13. Mystery**—winding paths, obscured features, flowing forms
4. Access to thermal and airflow variability—shade, radiant heat, seasonal vegetation		**14. Risk/peril**—floor to ceiling windows, water walks, high walkways
5. Presence of water—rivers, fountains, water walls, ponds, daylighted streams		

(continued)

Table 2.1 (continued)

Nature in the space incorporation of plants, water and animals into the built environment, especially with movement	***Natural analogues*** one degree of separation away from true nature; patterns and materials that evoke nature	***Nature of the space*** the way humans respond psychologically and physiologically to different spatial configurations
6. Dynamic and diffuse light—light from different angles, ambient diffuse lighting, circadian lighting		
7. Connection with natural systems—seasonal patterning, wildlife habitats, diurnal patterns		

Adapted from Ryan et al. (2014)

These design principles contribute to biophilic design as a new design theory on better contact with nature within and on buildings.

2.7 The Economics

The socio-psychological and environmental benefits previously discussed are likely to combine to contribute to significant economic benefits as set out in Fig. 2.1.

If humans are functioning better and their environment is working better, then the human economy is going to be more productive and efficient. While this is a logical assumption, it also needs to be backed up by data to be accepted.

Research has provided some quantifiable data that has enabled the economic case to be made, but it has tended to focus on the economics of either an individual benefit or a few connected benefits. The article '*The Economics of Biophilia*' (Browning et al. 2012), although still in grey literature, specifically focusses on 'making the economic case' for biophilic initiatives but restricts itself to the social benefits in workplaces,

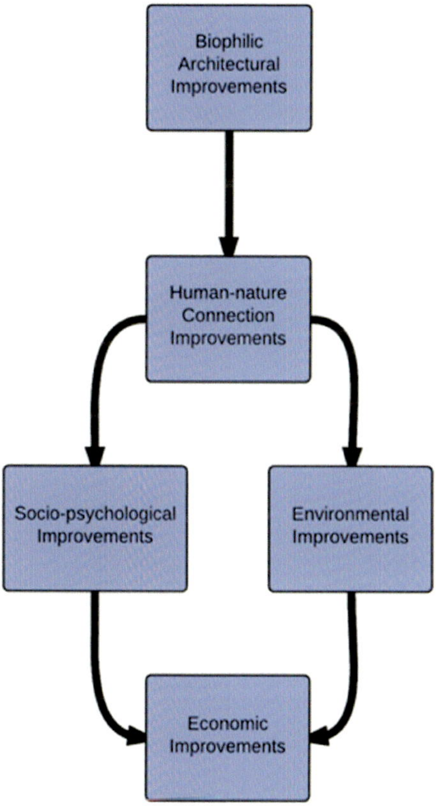

Fig. 2.1 Biophilic architectural benefit flow (*Source* Author)

health facilities, retail, schools, property value and crime reduction. With biophilia originating in the human–nature connection, this is justifiable, and Browning et al. support this further by pointing out that "today productivity costs are 112 times greater than energy costs in the workplace" and that daylighting schemes in offices can "save over $2000 per employee per year in office costs" (2012). The economic gains to be made from environmental benefits such as reduced energy costs, extended building life and decreased water management costs are apparent. Extrapolating the quantitative figures to support this is particular to location and local costs, but presents an area of research needing further attention.

2.8 The Multiple Benefits

Browning's paper was well received and timely, an easily understood document. It has helped overcome some of the perception that any biophilic design initiatives, particularly greenery, was an 'indulgent' extra. Cost-benefit analyses and continuous positive outcomes from case studies are consolidating the economic value of a biophilic design approach. Urban crises of heat and water management combined with increasing density are necessitating better urban design, where economics can no longer be the guiding principle.

The understanding and consideration of all the social benefits and all the environmental benefits is emerging, revealed through research and peoples experience of biophilic design initiatives. The multiple benefits are important as they make a strong business case for biophilic design implementation. Figure 2.2 lists the benefits in each category and illustrates how they combine to create solid economic benefits.

2.9 Analysis of the Literature

To deepen understanding of the uptake, interest and response to biophilic design, an analysis of the literature relating to the implementation of biophilic design principles was conducted. This analysis summarises the detailed review of the literature, some already presented in the chapter. Presenting it in a matrix (see Appendix A) with aggregated results reveals the numbers, trends and patterns, thus gaining indicative answers to the following questions:

- Which design principles are being investigated?
- What benefits are being identified in the research?
- Which aspect of biophilic design is receiving most attention?
- Where is the research being undertaken?

The selection criteria for the literature were:

- It linked directly to biophilic principles being put into practice.

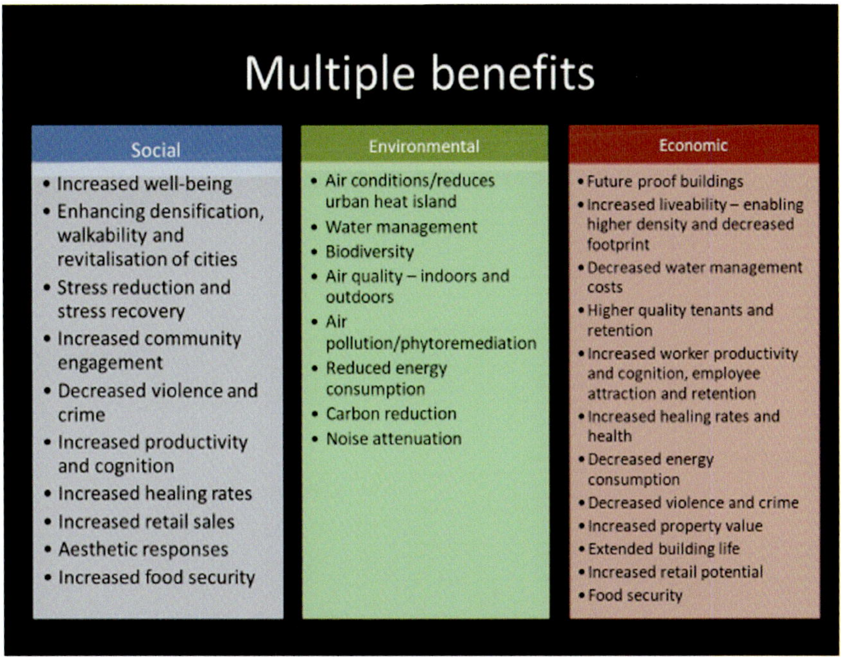

Fig. 2.2 The multiple benefits of biophilic design (*Source* Author)

- It was published in the last 10 years (unless an oft-cited seminal work).
- It was cited a minimum of two times by other authors.

X denotes area of primary research. O denotes secondary research.

The sorting of the biophilic literature into different aspects has been done using the categories of: social, environmental, economic and multiple (integrated analysis not clearly defined by one or other of these areas). These categories cover all the disciplines and professional practice areas that are used to discuss biophilics.

2.9.1 Summarised Outcomes of the Analysis

The first analysis (Appendix A.1) looked at design principles. The articles, design principles and which sustainability pillar they relate to are summarised in Table 2.2, with a graphical representation also in Figs. 2.3 and 2.4.

The analysis and figures clearly reveal that the biophilic design principle receiving most attention and research is the green roof. It also revealed that the research is focussing on the environmental benefits that can result from the implementation of biophilic initiatives. Green walls and indoor plants are the second highest design

Table 2.2 Biophilic design principles in research

		Number of articles
		47
Design principles		
	Green roofs	26
	Vertical living walls	10
	Indoor plants	10
	View of nature	3
	Green connectors	2
	Fractal patterns	2
	Natural lighting	1
	Bioretention basin	1
Sustainability pillars		
	Environmental	34 (32 Primary, 2 Secondary)
	Social	16 (11 Primary, 5 Secondary)
	Economic	10 (1 Primary, 9 Secondary)
	Multiple	8

Source Author

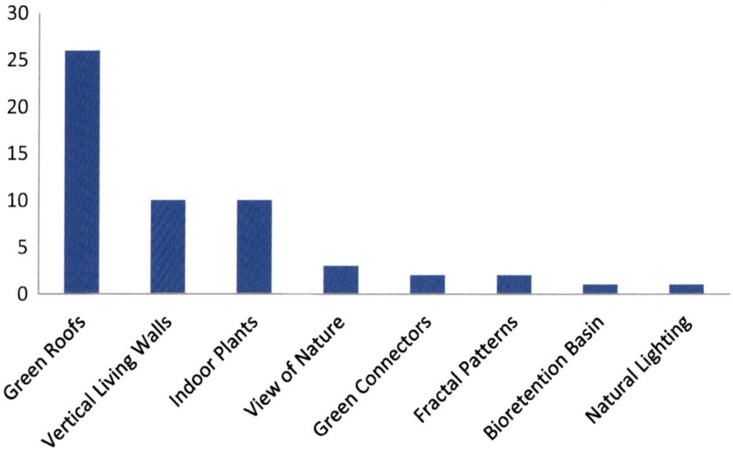

Fig. 2.3 Biophilic design principles in research articles (*Source* Author)

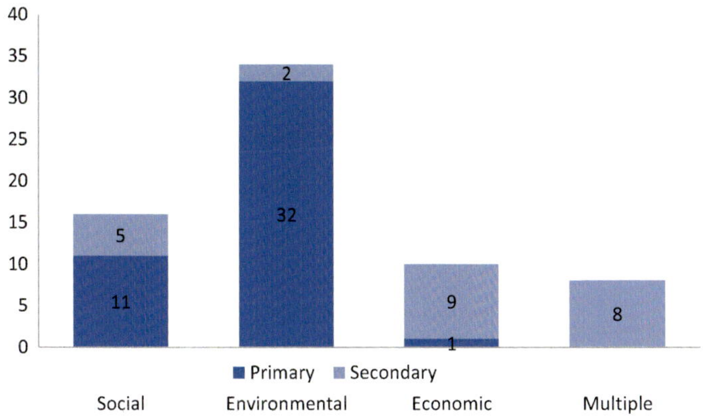

Fig. 2.4 Relationship to sustainability pillars (*Source* Author)

principle being researched. With green walls, the environmental outcomes are still the most researched but with indoor plants, it is the social benefits being considered.

This is significant as biophilic design originated through exploration of the human and nature connection. The fact that design principles are primarily being implemented for environmental reasons, not social, indicates that other factors are in play beyond the human–nature connection.

It was helpful to also consider from which countries the research was emerging. Countries such as Switzerland and Germany were not strongly represented in this review even though they are renowned for their traditional green roof implementation and technology. This is possibly because much of their literature is not in the English language. It is also possibly due to the fact that they have been implementing green roofs for some time and no longer need to research them as they are an accepted standard. Figure 2.5 and Table 2.3 depict from which countries the research is emerging.

The main countries from which the research is originating tend to be ones experiencing high urban density in their major cities. The larger share of the research came from the United States (US); the US is also where biophilic design as a concept originated.

Utilising the same criteria as the design principles literature, which aspects of the human responses to nature were attracting the most investigation and interest was analysed (Appendix A.2). Again, the results of the analysis were compiled in tables and graphs.

The analysis revealed that most interest was in investigating the benefits on human well-being which contact with nature can provide as shown in Table 2.4 and Fig. 2.6. Among the nine papers which investigated this broader picture, the reduction of stress was considered as a component of well-being. The next most frequent topic under prime investigation was stress reduction. This tends to indicate that stress is a significant issue in the contemporary world and that there must be baseline understanding that nature can help alleviate stress which triggers the initial research.

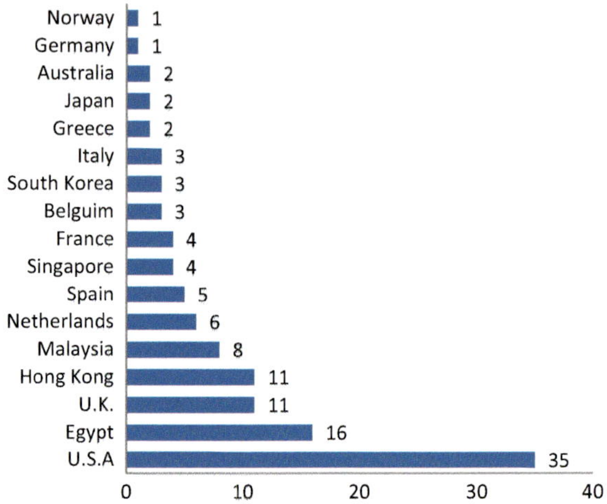

Fig. 2.5 Countries investigating design principles (*Source* Author)

Table 2.3 Countries investigating design principles

Design principles	
Country	Number of articles
U.S.A.	35
Egypt	16
U.K.	11
Hong Kong	11
Malaysia	8
Netherlands	6
Spain	5
Singapore	4
France	4
Belgium	3
South Korea	3
Italy	3
Greece	2
Japan	2
Australia	2
Germany	1
Norway	1

Source Author

Table 2.4 Most researched human–nature connection benefit

The human–nature connection in research		Number of articles 22
	Increased well-being	9
	Reduced stress	4
	Aesthetically pleasing	3
	Attention restoration	2
	Increased productivity	2
	Increased healing rate	2
	Cognitive enhancement	1
	Improved mental health	1
	Decreased violence and crime	1

Source Author

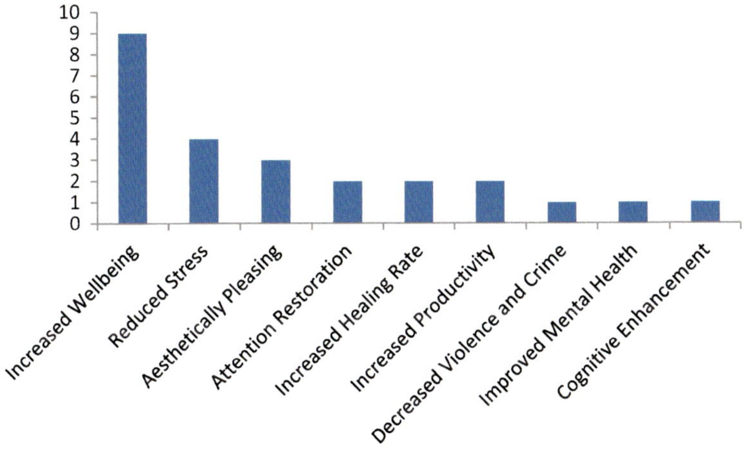

Fig. 2.6 Most researched human–nature connection benefit (*Source* Author)

It was interesting that aesthetics receives more attention in the research than healing rates or crime rates. The beauty of nature, though, has been extolled through various mediums for centuries.

Again, countries undertaking the research were considered.

The analysis revealed that a significant amount of research is occurring in the United States as depicted in Table 2.5 and Fig. 2.7. This was consistent with the previous tally of researched design principles. The other countries varied. The result confirmed that the US was a country active in implementing biophilic design principles and interested in researching the social and environmental benefits which result.

Table 2.5 Countries investigating human–nature benefits

Country	Number of articles
U.S.A.	10
Japan	5
U.K.	3
Australia	2
South Korea	1
Norway	1
Portugal	1
Sweden	1
Finland	1

Source Author

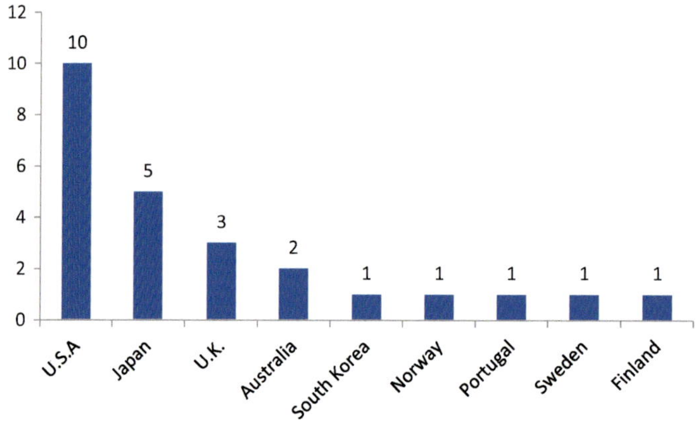

Fig. 2.7 Countries investigating human–nature benefits (*Source* Author)

2.10 Conclusion

Having further established the need for biophilic design within contemporary cities, this chapter then outlined the design elements that contribute to biophilic design. Once these were established, they were utilised to guide the review of the research investigating social, environmental and economic outcomes of biophilic design.

This revealed substantial growth in understanding of both psychological and physiological research on the human–nature connection in the last decade. Increased well-being and stress reduction were consistently revealed as areas in which there were positive responses to exposure to nature and for this reason were attracting a significant amount of investigation. There was good evidence for the environmental benefits which together with the social benefits translate into significant economic advantages. More research is needed in this area but biophilic design appears to be providing the theoretical foundation for the multiple benefits which can result from improving humans' daily connection with nature in the built environment. This mounting evidence of the human response to nature supports the theory of biophilic design as a valid, valuable and indeed essential approach to urban design and liveability.

The most investigated design principle is green roofs. Much of the research focused on the related storm water management benefits. Vertical green walls and indoor plants also attracted investigation. Aside from the benefit of storm water management, decreased energy consumption, indoor and urban cooling, air quality, stress reduction and aesthetics were the main benefits attracting attention.

The countries in which the research is occurring tend to be those with bigger, global cities where density and associated problems rather than the benefits have highlighted the need for more biophilic design. These cities were designed in an era of automobile and energy dependence, with a simple modernism that did not allow or encourage natural features in the built environment. Apart from recognition of the benefits of biophilic design motivating implementation, lack of space within such dense cities could be necessitating clever biophilic designs and utilisation of roofs and walls (Newman 2014).

Having gained an understanding of the benefits of biophilic initiatives as possible motivators and drivers for the expanding social movement of biophilic urbanism, a deeper journey of immersion into the social movement was required. Chapter three contains the stories of the pioneers and leaders in the social movement of biophilic design. They also contain a wealth of information on how to initiate a new approach, and how social movements are born and then progress.

References

Alexander C, Ishikawa S, Silverstein M (1977) A pattern language: towns, buildings, construction. Oxford University Press, New York

Appleton J (1975) The experience of landscape. Wiley, London

Beatley T (2011) Biophilic cities: integrating nature into urban design and planning. Island Press, Washington, DC

Berman MG, Jonides J, Kaplan S (2008) The cognitive benefits of interacting with nature. Psychol Sci 19(12):1207–1212. https://doi.org/10.2307/40064866

Berman MG, Kross E, Krpan KM, Askren MK, Burson A, Deldin PJ, Jonides J (2012) Interacting with nature improves cognition and affect for individuals with depression. J Affect Disord 140(3):300–305. https://doi.org/10.1016/j.jad.2012.03.012

Berto R (2005) Exposure to restorative environments helps restore attentional capacity. J Environ Psychol 25(3):249–259. https://doi.org/10.1016/j.jenvp.2005.07.001

Biederman I, Vessel EA (2006) Perceptual pleasure and the brain: a novel theory explains why the brain craves information and seeks it through the senses. Am Sci 94(3):247–253. https://doi.org/10.2307/27858773

Browning B, Garvin C, Ryan C, Kallianpurkar N, Labruto L, Watson S, Knop T (2012) The economics of biophilia—why designing with nature in mind makes financial sense. http://www.terrapinbrightgreen.com/report/economics-of-biophilia/

Burns G (2005) Naturally happy, naturally healthy: the role of the environment in well-being. In: Huppert FA, Baylis N, Keverne B (eds) The science of well-being. Oxford University Press, Oxford, UK, pp 405–434 (Reprinted from: 2007)

Guéguen N, Stefan J (2014) Green altruism: short immersion in natural green environments and helping behavior. Environ Behav 2014(July):1–19. https://doi.org/10.1177/0013916514536576

Hagerhall CM, Laike T, Taylor R, Küller M, Marcheschi E, Boydston C, Thwaites K (2012) Human EEG responses to exact and statistical fractal patterns. Paper presented at the 22nd international association people—environment studies (IAPS), Glasgow, UK

Hartig T, Bringslimark T, Patil GG (2008) Restorative environmental design: what, when, where and for whom? In: Kellert SR, Heerwagen JH, Mador ML (eds) Biophilic design: the theory, science, and practice of bringing buildings to life. Wiley, Hoboken, USA, pp 133–152

Heerwagen J, Gregory B (2008) Biophilia and sensory aesthetics. In: Kellert SR, Heerwagen JH, Mador ML (eds) Biophilic design: the theory, science, and practice of bringing buildings to life. Wiley, New Jersey, pp 227–241

Hes D, Du Plessis C (2015) Designing for hope: pathways to regenerative sustainability. Routledge, Oxon, UK

Ikei H, Komatsu M, Song CR, Himoro E, Miyazaki Y (2014) The physiological and psychological relaxing effects of viewing rose flowers in office workers. J Physiol Anthropol 33(6):1–5. https://doi.org/10.1186/1880-6805-33-6

Kaplan S (1995) The restorative benefits of nature: toward an integrative framework. J Environ Psychol 15(3):169–182. https://doi.org/10.1016/0272-4944(95)90001-2

Kaplan R, Kaplan S (1989) The experience of nature: a psychological perspective. Cambridge University Press, Cambridge

Kellert SR (2008a) Dimensions, elements, and attributes of biophilic design. In: Kellert SR, Heerwagen J, Mador ML (eds) Biophilic design. Wiley, New Jersey, US, pp 3–19

Kellert SR (2008b) Biophilia. In: Sven EJ, Brian F (eds) Encyclopedia of ecology. Academic Press, Oxford, pp 462–466

Kellert SR, Calabrese EF (2015) The practice of biophilic design. http://www.biophilic-design.com/

Kellert SR, Heerwagen J, Mador ML (eds) (2008) Biophilic design: the theory, science, and practice of bringing buildings to life. Wiley, Hoboken, NJ

Kuo FE (2001) Coping with poverty: impacts of environment and attention in the inner city. Environ Behav 33(1):5–34

Kuo FE, Sullivan WC (2001) Environment and crime in the inner city—does vegetation reduce crime? Environ Behav 33(3):343–367

Li Q, Otsuka T, Kobayashi M, Wakayama Y, Inagaki H, Katsumata M, Kagawa T (2011) Acute effects of walking in forest environments on cardiovascular and metabolic parameters. Eur J Appl Physiol 111(11):2845–2853. https://doi.org/10.1007/s00421-011-1918-z

Matsunaga K, Park BJ, Kobayashi H, Miyazaki Y (2011) Physiologically relaxing effect of a hospital rooftop forest on older women requiring care. J Am Geriatr Soc 59(11):2162–2163. https://doi.org/10.1111/j.1532-5415.2011.03651.x

Miller JR (2005) Biodiversity conservation and the extinction of experience. Trends Ecol Evol 20(8):430–434. https://doi.org/10.1016/j.tree.2005.05.013

Moore EO (1981) A prison environment's effect on health care service demands. J Environ Syst 11(1):17–34

Nanda U, Pati D, Ghamari H, Bajema R (2013) Lessons from neuroscience: form follows function, emotions follow form. Intell Build Int 5(1):61–78. https://doi.org/10.1080/17508975.2013.807767

Nesse RM, Williams GC (1995) Evolution and healing: the new science of darwinian medicine. Weidenfeld and Nicolson, London

Newman P, Matan A (2012) Human health and human mobility. Curr Opin Environ Sustain 4(4):420–426

Newman P, (2014) Biophilic urbanism: A case study on Singapore. Aust Plan 51(1):47–65. https://doi.org/10.1080/07293682.2013.790832

Nieuwenhuis M, Knight C, Postmes T, Haslam SA (2014) The relative benefits of green versus lean office space: three field experiments. J Exp Psychol Appl 20(3):199–214. https://doi.org/10.1037/xap0000024

Ode Å, Hagerhall CM, Sang N (2010) Analysing visual landscape complexity: theory and application. Landsc Res 35(1):111–131. https://doi.org/10.1080/01426390903414935

Orr D (1993) Love it or lose it: the coming biophilia revolution. In: Kellert SR, Wilson EO (eds) The biophilia hypothesis. Island Press, Washington, DC, USA, pp 415–440

Park SH, Mattson RH (2008) Effects of flowering and foliage plants in hospital rooms on patients recovering from abdominal surgery. HortTechnology 18(4):563–568

Park BJ, Tsunetsugu Y, Kasetani T, Kagawa T, Miyazaki Y (2010) The physiological effects of Shinrin-yoku (taking in the forest atmosphere or forest bathing): evidence from field experiments in 24 forests across Japan. Environ Health Prev Med 15(1):18–26. https://doi.org/10.1007/s12199-009-0086-9

Pert CB (1997) Molecules of emotion: why you feel the way you feel. Scribner, New York

Pert CB, Snyder SH (1973) Opiate receptor: demonstration in nervous tissue. Science 179(4077):1011–1014. https://doi.org/10.1126/science.179.4077.1011

Raanaas RK, Evensen KH, Rich D, Sjøstrøm G, Patil G (2011) Benefits of indoor plants on attention capacity in an office setting. J Environ Psychol 31(1):99–105. https://doi.org/10.1016/j.jenvp.2010.11.005

Ryan CO, Browning WO, Clancy JO, Andrews SL, Kallianpurkar NB (2014) Biophilic design patterns: emerging nature-based parameters for health and well-being in the built environment. ArchNet Int J Arch Res 8(2):62–76

Salingaros NA, Masden KG (2008) Neuroscience, the natural environment, and building design. In: Kellert SR, Heerwagen JH, Mador ML (eds) Biophilic design: the theory, science, and practice of bringing buildings to life. Wiley, New Jersey, pp 59–83

Söderlund J, Newman P (2015) Biophilic architecture: a review of the rationale and outcomes. AIMS Environ Sci 2(4):950–969. https://doi.org/10.3934/environsci.2015.4.950

Stamps AE (2002) Fractals, skylines, nature and beauty. Landsc Urban Plan 60(3):163–184. https://doi.org/10.1016/S0169-2046(02)00054-3

Tan PY (2013) Vertical garden city: Singapore. Straits Times Press Pte Ltd, Singapore

Taylor RP (2006) Reduction of physiological stress using fractal art and architecture. Leonardo 39(3):245–251

Tenngart Ivarsson C, Hagerhall CM (2008) The perceived restorativeness of gardens—assessing the restorativeness of a mixed built and natural scene type. Urban For Urban Green 7(2):107–118. https://doi.org/10.1016/j.ufug.2008.01.001

Tyrväinen L, Ojala A, Korpela K, Lanki T, Tsunetsugu Y, Kagawa T (2014) The influence of urban green environments on stress relief measures: a field experiment. J Environ Psychol 38(2014):1–9. https://doi.org/10.1016/j.jenvp.2013.12.005

Ulrich RS (1979) Visual landscapes and psychological well-being (National Parks). Landsc Res 4(1):17–23

Ulrich RS (1984) View through a window may influence recovery from surgery. Science 224(4647):420–421

Ulrich RS, Simons RF, Losito BD, Fiorito E, Miles MA, Zelson M (1991) Stress recovery during exposure to natural and urban environments. J Environ Psychol 11(3):201–230. https://doi.org/10.1016/S0272-4944(05)80184-7

Wilson EO (1984) Biophilia. Harvard University Press, Massachusetts

Wilson EO (2008) The nature of human nature. In: Kellert S, Heerwagen J, Mador ML (eds) Biophilic design. Wiley, New Jersey, pp 21–25

Yok TP, Chiang K, Chan D, Wong NH, Yu C, Tan A, Wong NC (2009) Vertical greenery for the tropics. National Parks Board and Building and Construction Authority, Singapore

Chapter 3
Biophilic Design: The Stories of the Pioneers of the Social Movement

Abstract Biophilic design qualifies as a social movement as it was driven by a group of people with a shared vision and passion to achieve common goals. The emergence of biophilic design was timely, the need was there plus the supporting research literature. A journey throughout North America was undertaken to investigate and understand how and why biophilic initiatives had occurred. Thirty biophilic design pioneers were interviewed and their stories written. Stories can be rich in information and insights, and through these the motivators and drivers of the biophilic design movement are exposed. The interview stories are also about passion and intuitive knowing, about local champions and how the right timing can lead to significant, and change-making, outcomes. The interviews also reveal strategies, policies and ideas to initiate and progress biophilic urbanism.

Keywords Heuristic inquiry · Social movement · History biophilic design · Biophilic pioneers · Biophilic Cities · Stephen Kellert · Tim Beatley

3.1 Introduction

The birthplace of the term biophilic design was at a gathering of people in Rhode Island in the United States. This chapter is the journey to North America, to investigate and understand what was driving the emerging social movement of biophilic design. It is told through stories collected through a series of interviews with 28 pioneers in ten cities of the United States and Canada. The stories contain a rich wealth of information, not just about biophilic design implementation but about people and passions. It concludes with two interviews held in my hometown of Fremantle, Western Australia.

Through these interviews, the aim was to investigate and identify what had motivated either an individual or a city to pursue the implementation of biophilic initiatives. The barriers were also identified. Each interview concludes with a personal reflection.

The journey began with a conference involving biophilic design and biophilic urbanism: the Biophilic Cities conference at the University of Virginia in October 2013. It was the inaugural Global Conference of Biophilic Cities, where the concept

J. Söderlund, *The Emergence of Biophilic Design*, Cities and Nature,
https://doi.org/10.1007/978-3-030-29813-5_3

was launched and discussed. This chapter begins with a discussion of my experience in attending this conference and any significant learning that came from this.

After the conference in Virginia, the next two and a half months were spent travelling to ten cities in the United States (US) and Canada and interviewing 28 participants. The conversation focused on their story of biophilic design: what their strategies and motivators were, the history of how things had evolved and what their thoughts were in regard to biophilic design. To gain an understanding of their connection to nature I asked whether they had grown up in nature or not. The answers would be recorded as to whether they had established a connection with nature while young or as an adult and whether this was with rural, wilderness or urban nature. I also asked for one word to describe their feelings when they were immersed in nature.

The interviews were open-ended in line with a heuristic inquiry framework, being allowed to flow with their natural form and time. They frequently felt like passionate conversations between two people with a shared interest. The interviews were recorded via audio recording, note taking and film.

Also included in the chapter are stories of two places I visited which demonstrate biophilic design principles and which contributed to the research by illustrating people's responses to the biophilic design principles in action.

In Fremantle, Western Australia outdoor living wall trials were conducted in conjunction with the City of Fremantle. The interviews with the Mayor of Fremantle and the contractor who installed the wall are included.

3.1.1 Defining the Parameters

Many design elements that are categorised under biophilic design by the biophilic design movement may not be put under that label by their originators and may be implemented for environmental reasons, not because of the biophilia human–nature connection. Biophilic design is expansive, encompassing many elements, so it was necessary to set clear parameters on what was being investigated. The focus narrowed to biophilic architecture with building-integrated vegetation such as green roofs and green walls, plus any new or innovative urban design initiative such as rain gardens and the daylighting of streams. This worked extremely well and provided the necessary boundaries in guiding the journey. In some interviews, the boundary was appropriately blurred between biophilic design and biophilic urbanism due to the intertwining of their stories.

3.2 Biophilic Cities: What Is One?

One of the latter chapters in the book *Biophilic Design: The Theory, Science, and Practice of Bringing Buildings to Life* (Kellert et al. 2008a, b) introduced the concept of biophilic urbanism and biophilic cities through the author Tim Beatley, the Teresa Heinz Professor of Sustainable Communities in the Department of Urban and Environmental Planning, School of Architecture at the University of Virginia. Beatley examined the inclusion of biophilia on different urban scales such as streets, neighbourhoods and communities, concluding with the vision of biophilic cities. Although labelled ambitious by Kellert in his introduction in *Biophilic Design* (Kellert et al. 2008a, b), Beatley pursued this vision with the writing of a book, *Biophilic Cities,* in 2011.

In the preface to his book *Biophilic Cities* (Beatley 2011), Beatley explained that the *Biophilic Design* book served to convince him "of the need for a stand-alone book about biophilic cities" (Beatley 2011, p. xi). In 2009, Beatley had already been involved in filming a documentary titled *The Nature of Cities,* portraying people around the world who are involved in projects focused on retaining our connection to nature within urban environments. In Beatley's online blurb promoting the movie, he suggests "It is our goal to raise the consciousness and understanding of this movement..." (*Biophilic Cities*, 2014). His book *Biophilic Cities* (2011) sought to expand upon the film and detail North American, European and Australian biophilic examples and stories. It explored the history of biophilia and included illustrations of biophilia that currently exist in cities. Beatley outlined the beginnings of defining a biophilic city with examples of current strategies and designs which progress biophilic initiatives in cities and concluded with future directions and what is needed in both research and practice. Beatley maintained a sense of questioning and curiosity throughout the book, recognising that it was only an introduction to characterising what a biophilic city is, what it looks and feels like. In his conclusion, he expressed how curiosity about the natural world by a city's residents may be essential and an indicator of the city's biophilic success. But the success needs to occur on all scales, from rooftop to region, and this entails a "significant popular shift in the way we view cities" (Beatley 2011, p. 151).

Beatley worked hard to progress his vision of expanding the movement and understanding of biophilic cities. A Biophilic Cities website was developed in 2012, incorporating a blog, resources, webinar links, news and a list of partner cities who have pledged to progress their biophilic qualities (http://biophiliccities.org/). The website, while offering ideas about what comprises a biophilic city, also opened up the discussion on what one may look like. The characteristics of a biophilic city were not defined, yet it was considered that the defining qualities are essential as without these there is the possibility of biophilic initiatives being greenwashed and shorthanded with tokenistic implementation measures (Beatley 2011, p. 156; Cramer and Browning 2008, p. 344).

To assist in defining a biophilic city, Beatley and his team at the University of Virginia were inspired to host a conference of delegates from the cities who had already become biophilic city partners. The cities shared the goals of stimulating urban biophilic initiatives and further defining of the expectations of a biophilic city. Also invited to the conference were key academics in the field such as Stephen Kellert and Jennifer Wolch, plus biophilic researchers and practitioners.

I was invited to attend due to my research, my trial living walls and as a representative of Perth, Western Australia.

3.3 The Inaugural Biophilic City Conference

Virginia—The 2013 Biophilic Cities conference

The buzz and excitement of this inaugural Biophilic Cities conference was contagious. Attendees had come from all across the globe to spend 3 days discovering and sharing thoughts, ideas, experiences and passions around biophilia in cities. Apart from coming from different countries, the attendees also came from different arenas. There were representatives from government, academia, civil society and the private sector, which included architects and green infrastructure consultants. Primarily, the attendees were from academia or government.

I presented a synopsis of Perth and of my trial green wall project with preliminary results of the social survey attached to the trial. The survey attracted much interest, especially the fact that, prior to analysis, initial responses showed people ranked the beauty of nature over the potential of nature to de-stress them. Kellert, in his keynote address, was to mention this survey result. It was surprising to discover that similar surveys were not known and some delegates were interested to replicate the Fremantle Greenskins survey in their countries. Beatley had also invited me to assist some UVA (University of Virginia) architecture students to build a small green wall and present another short talk about the process. The construction process provided an unexpected opportunity to experience the biophilic design principle of organised complexity (Fig. 3.1):

> The students had received instructions to plant the wall so as to 'look natural'. When I viewed their placement of plants, it looked contrived. So I themed the plants in their species and colours and placed them in clusters, running diagonally or patterned somehow. Interestingly, it then looked natural and was an immersive lesson for both myself and the students in the organised complexity of nature.
>
> (From my journal, October 2013)

Stephen Kellert and Jennifer Wolch from Berkeley were the keynote speakers, with Wolch as the first speaker. Her opening statement suggested that biophilic design is "a movement so powerful that it has exploded across the world" (Wolch, personal communication, October 19, 2013). This was exciting, as here was immediate confirmation that biophilic design was being perceived as an emerging movement. Wolch was unable to stay for the conference and so there was only a brief chance to talk with her. In asking what she considered was motivating this movement, she suggested that it was "climate change" which could have meant the widespread concern for cities

Fig. 3.1 Building a trial green wall. Beatley choosing a plant, bottom left (*Source* Author)

in a very uncertain and much hotter future as well as the multiple issues associated with burning fossil fuels.

All the Biophilic Cities Network partners from around the world presented an overview of biophilic initiatives occurring in their cities. The conference then concluded with another keynote speaker, this time Stephen Kellert:

> After outlining the history of biophilic design, he discussed the barriers to its implementation, introducing thought provoking concepts. He considers that humans have an experiential response to nature, reflective of our innate connection. These physical responses do not have to be relevant in our human evolution, but they are, and this is biophilia, evidenced also in non-human primates. Biophilia is a simple concept but it reveals itself in many ways, it describes our humanity. Kellert argues that biophilia is a weak biological tendency whose adaptive development depends on learning; it can be mal-adapted. Biophilia within people needs to be fostered, it needs social reinforcement.
>
> (From my journal; Kellert 2013)

This discussion highlighted my previous thoughts on whether the people driving the biophilic movement have had an upbringing in nature or significant exposure to nature at some time. Kellert raised another intriguing point in that we tend to underestimate the aesthetic value of nature, yet aesthetics, with fractal geometry and the organised complexity of nature, is a central issue of biophilia.

This was the first discussion of what would be a recurring theme in many of my forthcoming conversations in the US and Canada.

> Kellert mentioned the amount of increasing research revealing the benefits for including biophilic elements within urban design and the connectedness and sense of place that develops for people when nature is included in city design. Kellert posits that without this sense of place and emotional attachment, a building or city is not sustainable, even if it may be a Gold Certified LEED building. Through his experiences, he has recognised that a higher level of exposure to nature develops a deeper sense of care within people. Kellert suggested that sustainable biophilic urbanism and the growth of biophilic cities depends on a new consciousness towards nature rather than a new checklist. He concluded with the thought that "biophilia is a moral imperative".
>
> (From my journal; Kellert 2013)

The final day of the conference focused on identifying what was needed to progress the ideas and concepts expressed throughout, plus how to develop the Biophilic Cities Network. Participants gathered in groups to collectively brainstorm, some of the outcomes presented in Fig. 3.2.

We were first invited to identify the research questions, that is, what did we want to explore in this session and what do we want to focus on achieving? Large sheets of paper rapidly filled. For example:

- What is a biophilic city?
- What are the evidence-based arguments for the health benefits of biophilic urbanism?
- How do you measure a biophilic attribute in an individual or city context?
- How do you maintain initiatives or implement stewardship?

The lists were comprehensive and reflected the diversity of interests and knowledge. The barriers to biophilic design initiatives were then discussed with common themes such as

- Lack of knowledge,
- Fear of change and innovation,
- Lack of leadership,
- Lack of political will,
- Short-sighted thinking and
- Financial mechanisms do not reflect the benefits, especially in health, of biophilic initiatives.

These were barriers encountered by this global group in their fields such as depicted in Fig. 3.3.

To conclude, the groups outlined goals that were perceived as helpful to the progression of biophilic urbanism. These included

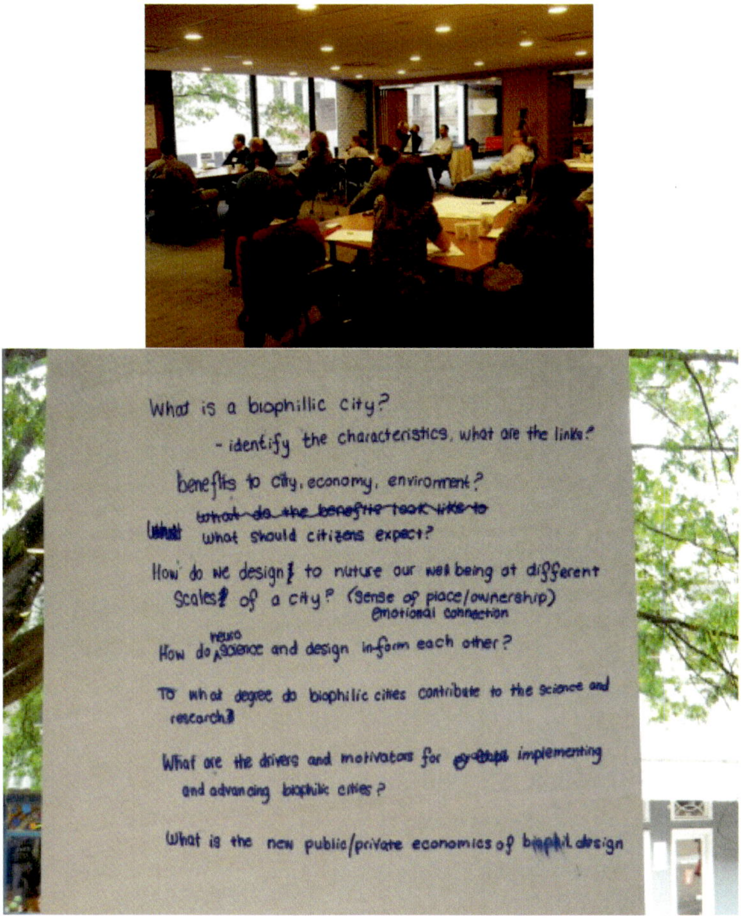

Fig. 3.2 The biophilic city discussion (*Source* Author)

- Cultivate a biophilic consciousness,
- Cultivate a practice of awareness,
- Showcase best practices,
- Advocacy and adoption,
- Networking and
- Strategies for communicating the urgency to leaders and communicating the multiple benefits on offer with biophilic initiatives in urbanism.

Throughout the conference there appeared a shared understanding, an intuitive knowing that did not need to be spoken. There was a common language, a shared goal, a shared vision and an understated but apparent connection to nature. The appreciation of the beauty of nature encountered on field trips was expressed. Deep conversations were held and a sense of family developed even though participants

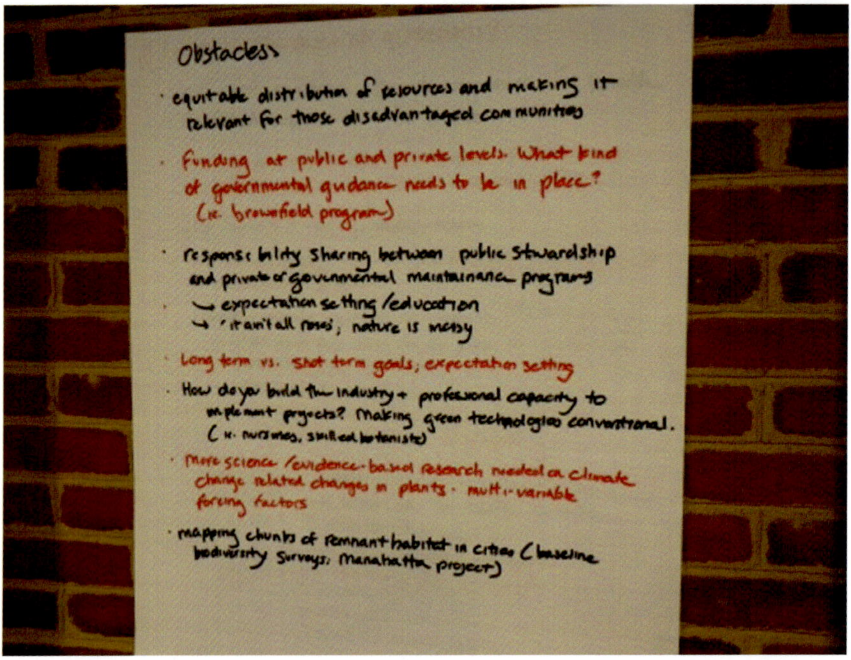

Fig. 3.3 The obstacles as identified in the discussion (*Source* Author)

were from different backgrounds and different countries. There was unity under the collective identity of biophilic design.

3.4 The Biophilic Design Pioneers

At the conclusion of the conference, planning begun for the ongoing journey and the compiling of a list of key players and pioneers in the social movement of biophilic design. The conference was an excellent beginning as many of the attendees lived in North American cities and had connections which they were willing to share as well as participate in an interview themselves.

3.4.1 The Core Pioneers

Three of the participants at the conference had attended the Rhode Island conference and contributed to the emergent book *Biophilic Design*. All agreed to be interviewed when I travelled through their city and they could also connect me with another pioneer, Judith Heerwagen, when I was in Seattle.

Tim Beatley—Virginia—29/10/2013
Teresa Heinz Professor of Sustainable Communities
School of Architecture
University of Virginia

After the conference, Beatley and I were able to have an extended conversation about biophilic design and its progression since biophilia was first popularised by Wilson. Beatley's previous career focus has been in green urbanism with the contemporary language framed by technical and metabolic values. Biophilic design was different for him as "biophilia brings thinking in about how these things affect humans. A low ecological footprint energy efficient building may not be nourishing or even necessarily pleasant to be in".

Beatley and I discussed the concept of biophilia being a weak biological tendency that Kellert had introduced in the conference. We both agreed that greater exposure to nature does nurture a connection. Beatley suggested the tendency "is weak in the sense that like a muscle, it needs to be exercised". This led to a discussion on whether people who are passionate about nature have also spent a lot of time in nature. Beatley pointed out that many of the great contributors in the environmental field such as Rachel Carson "have had some real important childhood upbringing that involved significant contact with nature. So there is certainly something there. Your exposure at an early age predisposes you to understanding the value and importance of natural contact".

As an urbanist and an urban planner, Beatley thinks the challenge for him is to look at the whole cumulative elements of nature in a city, the green networks, the bird sounds and the wildlife. "It is these cumulative effects of all of nature that triggers biophysical and emotional responses to something". Biophilia provides a different framework; it introduces our emotional and biological needs. As Beatley suggests, "it's a different approach to just protecting nature as another urban 'infrastructure', just fixing another engineering problem".

Beatley thinks how we connect people to what is around them through social and organisational structures is the challenge. "We need to be able to build on this natural, social capital of connecting people to the nature around them. There is a lot of nature in cities that is not very visible, or very small or not understood. How do we get visible that nature and get kids interested?" Much of the environmental policy, Beatley thinks, "has been driven by the belief that market systems fail in the recognition of the need for nature in our cities". The perceived need for benefit/cost analysis, ecosystem analysis or economic analysis is a "slippery slope" that will influence conventional thinking. There is a need for this but Beatley resists the idea "that economics becomes the driving metric. If this is the sole criteria then you live and die by it. If a green roof does not deliver the expected temperature reduction and energy savings then the policy can go away".

I ask Beatley if he thinks there is a social movement happening in biophilic urban design. He replies "there is a new movement in the way people are talking about nature in our cities, and who is talking. The dialogue now involves people from city planning, architecture, urban design, public governance and public health.

There has been a rediscovery of health as a driver, brought on by the obesity crisis along with stress and mental health issues". There are "new pushes, new dialogues, new discussions". Parallel to this, Beatley explains, has been Richard Louv's book *Last Child in the Wild* and "the recognition of the profound disconnect that has happened in my lifetime for children from nature and the outdoors. This is new; the disappearance of the 'free range child'". Louv's book, in Beatley's experience, has ignited something and touched a chord. At the same time, the food movement has grown, food security has become an issue, so there is a "renewed appreciation of the importance of being able to grow food".

Beatley has an awareness of certain things converging. "It can be health or community or connection to nature. This is what is new. Some of these are old values, old ways of common sense that have been dropped and are being revisited with a new appreciation and vigour. Biophilia touches them all".

Beatley described how in Birmingham they are looking at a whole variety of issues such as stress, heat, drought, climate change and poverty that contribute to climate stress and layering all these things. They are then "imagining how the whole natural system of that city could act as a way of addressing this urban stress and tackling all these issues".

We went back to discussing the movement we had both recognised was occurring. Beatley had been at the Rhode Island conference so I asked him about his experience and thoughts and how biophilic design emerged. "A lot of people had been working in this area and calling it different things. Architects doing organic stuff with natural materials didn't call themselves biophilic designers but Steve got them together and we started using that word. These were different professions doing different work and gradually moving in the direction of the language of biophilic design. So it's probably only been about a decade that biophilic design has been used. But I haven't heard of Biophilic Cities used before".

The discussion moved to Chicago and the implementation of the Chicago City Hall green roof by Mayor Daley. We talked about whether research would find initiatives were being introduced top-down or bottom-up. Beatley thought it is a blend of both: "There is a lot of bubbling up from the bottom and there is a lot of urban greening stuff ending up as a subsidy from top down". We concluded that there are a lot of good stories to be collected and recounted that would capture a sense of what is happening and why in the biophilic design area in the US.

Beatley described the house where he grew up: "I had an ideal house and neighbourhood, essentially grew up in a single house (until I went off to college) that my father built on 6 acres of land, but within a city. It was the best of both worlds" (Fig. 3.4).

The interview with Beatley was the perfect beginning. It was a great overview of biophilic urbanism. Beatley was able to express his thoughts on what he understood was happening, particularly in the US. He agreed there was 'something' different going on in the way urban issues were being talked about and addressed. These issues were bringing together people from varying disciplines and biophilic design was providing a framework, as it had at the Rhode Island conference.

Fig. 3.4 Beatley's house his father built (*Image courtesy* Tim Beatley)

Stephen Kellert—Yale—11/11/2013
Tweedy Ordway Professor Emeritus
Yale University
School Forestry & Environmental Studies

Having met me at the Virginia conference, Kellert agreed to an interview and this was conducted by phone while I was in New York. In 1992, Kellert alongside E.O. Wilson brought people together to explore the concept of biophilia following the interest in Wilson's book, *Biophilia*. It was from this gathering that the biophilia hypothesis developed and the subsequent publication of *The Biophilia Hypothesis* (Kellert and Wilson 1993). "The basic premise was that humans have an innate, genetically hardwired connection to nature that is responsible for psychological and physiological responses to nature and the values and patterns of nature. Aesthetics is one of these values".

Kellert discussed what he considers as our inherent tendency and need to learn. "There are obstacles to learning and one is our modern design and aesthetics. It disconnects us from the natural world. This is why biophilia is needed as a design concept".

The development of biophilic design occurred at the 2008 Rhode Island conference which was spearheaded by himself, Bill Browning and Judith Heerwagen. "The conference brought together different people and professions whose work touched on the biophilia concept". It was from this conference that biophilic design evolved into a published book.

In Kellert's lecture at the Biophilic Cities conference and in this interview, he stressed the importance of aesthetics as inherent in biophilic design. "Beauty has a sound evolutionary adaptive function. It may be distorted in design, or considered a luxury rather than a necessity, but is a basic concept of biophilic design. We need to connect people to nature for them to be happy and whole. Incorporating a superficial nature feature in a building does not constitute biophilic design. It needs to connect people with nature".

Kellert considers there is not a prime motivator to designing biophilically: "The multiple benefits that can ensue from biophilic building provide a list of potential motivators and drivers, each of them important. Small spaces designed in the right manner can provide places of beauty and wonder and provide children with an opportunity to experience this".

Kellert became interested in wildlife and nature as a college teenager, immersing himself as much as he could in nature. He considers that this connection can be awakened or triggered at any stage of life. "Nature is a part of humanity's feelings, to cut off from the natural world is to cut off from part of humanity" (Fig. 3.5).

Further encounters
Since these initial meetings with Stephen Kellert I have had the opportunity to spend time with him in Singapore at biophilic design workshops for intended developments, hosted by the Singapore government. Tim Beatley was also present. I include these further encounters as they were informative and, as Kellert is no longer with us, important to record.

Fig. 3.5 The author and Kellert at the Biophilic Cities conference (*Source* Author)

In an introductory lecture, Kellert defined biophilic design as being *ecological*, with a sum greater than the parts; *experiential* and *emotional* as it is a fundamental emotional driver; *evolutionary*, and *validated* as it is recognised and can be related to. Biophilic design can be modified and adapted to varying spatial scales. It promotes stewardship as designs evoke memory and culture. Our built environment is our natural environment, so designs have to involve more than one profession, with economics not being a value, but a valuation strategy. The value of biophilic design needs to be captured so biophilic design metrics are needed, and for multiple climates. Cities ought to be viewed as Arks.

Kellert provided feedback on the workshopped biophilic concept plans for the developments and his wisdom really highlighted the potential of what a biophilic approach can bring. He often mentioned the need for seamless integration between the surrounding, or natural environment, and the built environment. It was easy to vision developments where the patterns, forms, colours, etc. of mangroves and mudflats, for example, could flow into the design of the built environment and infrastructure.

The hospital visit

Khoo Teck Puat (KTP) is a well-known and well-regarded hospital in Singapore which has been integrated and designed with biophilic elements, resulting in positive outcomes and great rewards. Following on the KTP success another hospital was built incorporating even more biophilic design elements: Ng Teng Fong General Hospital (NTFGH) and Jurong Community Hospital. The two parts linked by beautiful healing gardens (Fig. 3.6).

We had the honour to be taken on a tour by the hospital's director before the fully completed hospital was open to the public. The director discussed how all around the hospital there was life, to promote living. The hospital was full of organic shapes and plants, with good ventilation coupled with the ability to open louvres. These features were to promote healing (Fig. 3.7).

Light was considered important to help biorhythms and human's circadian clock. The design was based on research and the KTP hospital results. Biophilic design was an overarching design principle with initiatives also based on salutogenic principles. These principles are complementary and overlapping as both focus on factors that improve human health and well-being. In this regard, a waterfall which was created in one of the hospital's curved building canyons was considered important not just visually but also for the sound of water falling (Fig. 3.8).

Through clever herringboned patient bed configuration, every patient gets a window view. The hospital has balconies, podiums and walkways with greenery and gardens to assist with these views and the sounds of nature from the bed, but also so there is access during recovery. These initiatives are significant in reducing staff stress by providing quiet and beautiful spaces for staff to take some time. There is nature artwork, some of it the staff's, and many natural materials throughout the hospital. At one point, Kellert and myself are looking out an upper level window and he pointed out the ledge that ran around outside the floor to ceiling window. "Here", he said, "we have a great example of prospect and refuge but also demonstrating the need for this transitional space. Because of the ledge, people cannot look straight

Fig. 3.6 NTFGH model (*Source* Author)

down and be disturbed by the height. They look straight out and enjoy the view" (Fig. 3.9).

The taxi ride

On the way back to the hotel from the hospital, I had some precious time to talk further with Kellert. He told me about his 'epiphany', his moment of understanding about the innate and evolutionary perspective of human–nature responses. It happened while immersed in the Newfoundland wilderness, which, we both agreed, is probably why. This story gives greater detail and depth to the story of the emergence of biophilic design.

This is how he told it:

> He just realised that the values of biophilia he was exploring were genetic, and evolutionary, and he needed to create an affirmative analysis. So, there in Newfoundland, he called his colleague, Scott McVay, to talk about getting a group together to discuss a biophilia hypothesis. Kellert recognised that they needed the systemic inquiry and scientific conversation that a hypothesis brings.
>
> Kellert said he had first met Edward Wilson, who wrote *Biophilia* in 1982 so had been thinking about biophilia for a few years. With McVay, they were able to secure some funding and brought together an exceptional group of "very bright, high powered people. The group was contained and focussed in a natural setting with no audience to perform to."
>
> They agreed that biophilia had to be ecological, evolutionary and necessary. And also that it was wholistic and emergent so there wasn't a check list that would work, They needed

Fig. 3.7 Curves, canyons, greenery and louvres (*Source* Author)

> to connect pieces to each other in an integrated sense. This led to the book *The Biophilia Hypothesis*.

Our taxi ride finished, and Kellert invited me to call him when we were both home to continue the discussion and the story. I delayed the call and missed this opportunity as Kellert passed away a short while after.

Spending these times with Kellert was a privilege. He had a generosity of spirit in sharing his knowledge, with a keen observation of details. We found a common bond in our love and connection to nature: our biophilia. There seemed to be a shared understanding of how many of the world's stresses could be alleviated through greater integration of nature in daily lives.

Judith Heerwagen—Seattle—06/12/2013

HPGB Program Expert
Office of Federal High-Performance
Green Buildings
US General Services Administration

Affiliate Assistant Professor
University of Washington
College of Built Environments
Department of Architecture

Fig. 3.8 A water feature in a canyon provides healing sounds to patients (*Source* Author)

Heerwagen and I began the conversation agreeing that much was happening in the biophilic design movement. She mentioned that *Biophilic Design* has increasing sales. This book, she feels, was an impetus for a lot of the current interest in biophilic design, but it didn't happen right away. "It's been a gradual thing that as more people become interested in it, it sparks a little bit of competition to know more and do more, use it. I think that is part of it". I explained that my research so far had uncovered definite themes in what is motivating this interest and Heerwagen responded:

> The thing that is really interesting is how inter-disciplinary this is. It is cutting across design professions and academics' interest. You have landscape architects, psychologists, architects, urban designers. This is one of the more interesting aspects of this. It is not led by a particular profession and it is not clear who the key spokespeople are. They come from different backgrounds and are looking at it in very different ways. There are so many ways to interpret what biophilia means and how to do it. I think this makes it a very rich design palette. There is no standard that says this is how you should do it.

Heerwagen says there are people who are saying how biophilia should be a credit with the LEED rating system. She argues "How do you know if something is biophilic, you have to have some way of saying that you need to do this or that, and I find that very hard to do. I have been asked in the past to look at what a biophilia credit would include but I don't think you can use it in an abstract way; you can use it in a literal way. There are so many different ways to interpret what it is and how to do it. It is not just a green roof or not just a garden". Heerwagen said it is about design of the

Fig. 3.9 Kellert discussing biophilic principles regarding the ledge viewed through the window (*Source* Author)

space and utilising things like the sense of prospect and refuge: long-distance views from a sense of protection. "They don't need to be green nature spaces". Prospect and refuge were exemplified in the concrete building we were meeting in and Heerwagen pointed this out, also commenting that some people would not think of it as biophilia. She does think the most powerful elements are the natural ones: the greenery as well as sensory fluctuations. "If there was a vase of flowers on our table it would make this meeting feel very different. And there is research on flowers that, believe it or not, cause you to make a smile that engages your eyes and you know it is a true smile because of the muscles it engages in your face". Heerwagen explains that people don't express this same smile when they are given an orange. She considers there is something about flowers that is deeply important to us and that they make us happy. Heerwagen explains that this has a very good biological reason due to our evolutionary past where flowers were a sign of resources in our future; they produce nuts and fruits that we can eat. We discussed the aspects of aesthetics then, why do we find flowers pretty, or particular views? Heerwagen explains:

> Things are inherently beautiful themselves and they are beautiful to us because they present things that are useful to us. We like views of a golf course. It gives us a long distance view. And if we look at this from an evolutionary psychology perspective, that really is telling us about time and space. We can see things that are coming over the horizon. We can see a storm coming, or a dangerous animal. We can see in advance and plan for it, so there is a

real evolutionary advantage to be able to see and not be seen, to feel protected. These are the sort of views you find on calendars.

Where people find things beautiful, Heerwagen always looks for the reason. She suggests that people find things beautiful because they are useful, and beauty attracts us to things. So innately we are attracted to useful things that may aid survival. We discussed Paley Park in New York which I had found beautiful, agreeing it is a special place and that it offers the sense of refuge and enclosure and protection. Heerwagen says that she thinks these features are particularly powerful there.

I mentioned my interest in improving mental health through biophilic design in prisons and Heerwagen had read some research which had revealed that prisoners with a view had better health. The researcher was influenced by Ulrich's studies of hospital healing rates and wondered if this could also occur in a prison.

I brought up my perception so far of there being multidisciplinary drivers for biophilic design, and of them not necessarily being 'biophilic' per se but often as a response to a city's crisis. Heerwagen responded:

It is not driven by biophilia but I think people say that if we need to do this, can we do it in a way that is not ugly? That actually adds something nice to the city. Using a swale rather than a concrete surface. Any time you can gain multiple benefits with one approach, I think you have really gained a lot. Biophilia can be used to enhance the attractiveness of features that otherwise would be very ugly. Roofs are very ugly. When you are in a building a lot of times you look out at a roof and the HVAC system. Using a green roof for rainwater purposes as well as enhancing the appearance and beauty of a place; I think there is the sense that if you can do it for one reason, why not do it for more and design it in a way that has multiple benefits. In terms of the drivers, I think for buildings anyway, a lot driving this are actually the design firms. They have not been very interested in this for a long time and I think they are beginning to see it as a competitive advantage. There are a couple of very large firms in the US who are beginning to say: How can we do this? How can we integrate biophilia into our approaches? How can we do this in a strategic way? How can we really think about our designs through this lens? Everyone knows them and I think once they begin to do this….

Heerwagen then explains what happened with LEED and the sustainability movement was that cities began to require LEED certification for their buildings, not all but enough, and architecture firms began to realise they were at a competitive disadvantage if they didn't know LEED requirements. So they started getting their people trained and then the LEED system took off very dramatically once that happened.

And I really think that's going to happen with biophilic design. I think there are several big firms that are beginning to do this. They design buildings that get a lot of attention. Once buildings get a lot of attention then clients are going to want to have their folks design them. And if you can get them to integrate this with sustainability, most of the big firms now really take this seriously, then you get the double benefits. I hate to say it but a lot of the LEED buildings are not very attractive for the people. I am focussed on the people. Sustainability is a human benefit, biophilia is a human benefit. A lot of the LEED buildings are just focussed on technology and energy and water savings. A current topic of discussion is whether sustainability is creating buildings that are so technological and kind of grey in terms of people; that we have gone too far in the technology without paying attention to the human component and biophilia is a way then to begin to address the human factors. I think it is a natural component of sustainability that has not been adequately recognised. It is not just about technology.

Heerwagen has done evaluations of several very high-level LEED buildings, certified as Gold or Platinum, and some, she comments, have had about the worst ratings for comfort and satisfaction she has ever seen in buildings from a human perspective. One was described by an occupant as looking like a beautiful corpse. It didn't have any sense of life or something that softened the edges of technology. Figuring out how to integrate, and how to do the biophilia is the key. Indoor plants may be problematic, so how do we do the biophilia really well? How do we capture the essence of biophilia? She thinks Bill Browning has some good ideas for this. We can use fractal patterns, light and colour, design in abstracted ways. This should be looked at, not as an afterthought but as an integral part of the design from the beginning. "We are going to do sustainability, but we are going to do biophilic sustainability. I think you get a very different approach and a very different attitude if you start right from the start. Not use it to beautify your building but use it as an integral part of your strategy".

Heerwagen thinks there are lots of approaches and the palette is virtually unlimited in terms of using some of the already standard features such as colour, pattern, texture and light. The idea is to use it all in a more biophilic way that is derived from natural images and imagery. What impact does nature have on us and how do we integrate those positive benefits into a building? She sees that an understanding of biology is needed as well as an understanding of design. "Biologists don't understand building design and designers don't have the biological background. It's really hard to make the connection. So there are not many examples at the building level, it is more at the product level. Larger scale is harder". Traditionally, landscape architecture has been about making the building look good after construction, but now it is positioning the plants so it looks good from inside the building. Landscape architects need to come in right at the beginning.

Heerwagen can understand that sometimes there is a 'push back' from people who have been implementing biophilic initiatives such as green roofs purely for function for some time. Now it comes under a 'new-fangled term'. There are others asking "Why should we invest in this? What are the economic advantages?" This is always an issue too. Heerwagen questions if decision-makers who invest are paying attention to the economics. These are the ones needing to be influenced. Heerwagen wonders if Browning's paper *The Economics of Biophilia* was put in front of decision-makers whether they would believe it. Developing a believable case with good examples and stories would help. Even sustainability is still a hard sell in some areas because people think it may cost more, which it doesn't have to. This is why she thinks biophilia may get a push back in areas. It is just another new idea the design world is trying to sell. It needs to be integrated into policy and people in different areas need to talk. The benefits cross over into different fields and recognition and integration of this need to happen.

Heerwagen thinks to some extent there is a problem with the term biophilia. "It sounds like a fad. It is a much more serious topic than it seems to be with this word". She considers E.O. Wilson's book was wonderful, though, as the intention was to really look at how connected we are to the natural world.

The benefits of our connection with nature can be explored across age groups, especially with children. The benefits children get from connection with the natural world as opposed to a sterile, urban environment have been looked at. This can be explored further through the life cycle from young right through to the elderly. She thinks that there should be a greater focus on the benefits to people. These benefits are so powerful and can be followed through the life cycle. Heerwagen really wishes we could call this something other than biophilic design; it really is this intrinsic connection to nature that is so powerful. There are so many ways to touch on this. "The human benefit recognition is what is really needed to pull it together in a way that everybody will understand". At this point, policymakers need to decide that we just have to do this. It hasn't been integrated. The benefits may be different for everybody.

We discussed how people are willing to pay more for views of green. Heerwagen intriguingly suggests that this really is illogical. Why pay more for exactly the same apartment with a view when you can pay less?

> In New York they also pay for views of the sky. If people wanted to maximise their utility they would find a way of doing this the least expensive way as possible. And we know they don't do this. They are buying pleasure. They are buying prestige and things that have nothing to do with the basic utility and functionality of the apartment. A lot of the biophilia is about the hedonic value of space. It is illogical, if we were maximising our cost benefit ratio we would never do this. Which shows the real power of biophilia, quite frankly. We will pay more and there is good economic evidence, so we can see there is value other than money.

I recounted the Chicago Michigan Mile story (refer Michael Berkshire interview) of people paying more for views of a green roof over lake views and it is these stories that Heerwagen says need to get out there. "There was economic value to the building owners but hedonic value to the tenants". Heerwagen talked about the huge development that the company Amazon is doing in Seattle. Land is so valuable that Amazon is occupying all they can with buildings and losing the opportunity to put in any green amenities or even more street trees, anything that enhances the naturalness of the space.[1]

Heerwagen mentioned a story of when she had been giving a presentation on biophilic design to a big design company. She had been presenting the accumulating evidence and one woman asked "Why do we need to study this? We already know this?" Heerwagen responded "You are absolutely right but there are a lot of people who don't believe it until you do provide the evidence". The woman replied "But we know this intuitively! We know these things are good for us, we are attracted to these things, so why are we spending all this money on research when we should be spending it on doing it?" Heerwagen thought this was a good point. This is the gap, between researchers wanting to prove it and designers just wanting to do it. "So the question is why do we need the evidence? But you always have designers who are trying to prove to their clients that this is worth doing and the only way they are going to do that is to show that it has economic or linked economic benefits. So the

[1] Amazon has since built three glass domes which serve as an employee lounge and workspace.

design world is in need of the evidence at the same time they are just wanting to do it. If they had their way a lot of them would be doing it much more but they have to convince their clients".

Some investors respond with a "let's just do it" and others say "this is nuts". But Heerwagen doesn't think this should be seen as something different. It shouldn't be that we want to do this biophilic thing; it should be that this is the way we design. It is as if biophilic design is talked about as an add-on rather than an integral part of what is just done. She thinks people do understand that having a pleasing place is better than having a place that turns people off, or they don't want to come to or stay in. Heerwagen thinks most clients would recognise this and that biophilia should not even be mentioned. "They don't need to say 'we want to do biophilic design', we want to do great design. And biophilia should be part of the pallet of what great design is". Somehow biophilia has been seen as an add-on and it should never be seen this way. Heerwagen recounted a story to illustrate this point:

> A Federal building was being constructed on the site of a river that had been modified by Army engineers to become a canal. The designers wanted to connect the building to the original landscape, so they looked at what the river used to be and used this as their theme. This became their mission. They would have liked to have had an interior stream through the building, but compromised with running a stony river bed throughout with patterns of a winding river on the glass. They have put some trees in there. The concept was never questioned, no one said it would cost too much. The building has won design awards.
>
> The same design firm has also done some work with a children's hospital taking a similar approach. They have used themes of nature and animals. It was not sold as anything different. Hospitals are stressful places so it was about answering what can we do? How can we create a place that is interesting and pleasant and makes you want to be here? This is a firm that is taking this stuff really seriously and again it becomes part of how they design. Not an add-on, not a biophilic add-on.

Modern architecture closes off buildings, making them sterile with the emphasis on function rather than people. The building is a technological feat, not a habitat. I described the many building designs I had seen in my world travels, in the less developed countries. These were designed for function and to be practical but I had found them very beautiful as well. Heerwagen suggests that this was because they were very connected to the natural environment, conveying a sense of being in nature rather than apart from nature. Cities do not convey this same sense, but interiors could very much be designed in a different way.

Our discussion then diverged to why some people are more attracted to nature than others, or seem to need it more. Heerwagen also thinks it is to do with where you grow up. She spent all her youth outdoors, most of the day was outside. "Being inside was torture for me. This is part of the problem with young people now. They don't get the opportunity to be challenged by being outside. I would think people with an interest in biophilia would have had this connection with nature. I think there is an interest from young designers in this topic. Things like the farmers markets can connect them". We discussed Seattle and Portland and their easier connection to nature than places like New York.

Again our flowing conversation took a digression as I mentioned some recurring themes from previous interviews, such as the links between creativity, the open mind, connectivity, social innovation, self-organising systems, all helping build social capital. Heerwagen jumped on this asking how do we connect this to biophilia? She thinks it is an intriguing idea that makes a whole lot of sense. "Can we design biophilia in a way that actually enhances social connectivity? Most of it is thought of at an individual experiential level. Can biophilia be used to enhance the social aspects of space? If we had the biophilic Rhode Island conference in the middle of New York, would it have been the same? Even though we didn't go outside a lot I intuitively think that it would have been very different in how people interacted and the ideas that they got. I haven't got any proof of this but being surrounded by forest rather than inside a hotel must have made a difference. There is evidence for the fact that if you are in a positive mood you are more likely to be creative. Creative discussion is a really important part of something like that".

Heerwagen thought it would be interesting to research the difference between two groups given the same task, one group put in nature and the other a windowless building. Heerwagen laughs: "But then why do we have to prove this when we already intuitively know it!" I mentioned that a recurring theme had been the need to tap into this intuitive knowing, to put aside assumptions and approach things with an open mind. Immersion in nature can aid this. Another strongly recurring theme in the interviews has been aesthetics. Heerwagen agrees.

When I asked for her one word to describe her feeling she has in nature, her word was 'calming', but she also finds she de-stresses, she doesn't think when she is in nature. "I just enjoy the sights and sounds and I am not ruminating about anything, I am just experiencing it. There is a calmness and a sense of emptying the mind. There is a sense of relief. It is physiological as well as psychological" (Fig. 3.10).

The time spent with Heerwagen was a highlight of the journey. I am sure we could have talked for hours, bouncing ideas and sharing thoughts. It was wonderful to be able to learn so much in a few hours from a person so rich in experience and knowledge. Seattle, and this interview, was towards the end of the journey, so I was also able to share previous understandings and experiences from other cities, and people we both knew.

Seattle was particularly fascinating as it was dealing with unique issues. Big companies such as Amazon and Microsoft have created particular issues around building design, urban space and density. It is not only the headquarters but also housing for many employees.

Bill Browning—New York—13/11/2013
Director
Terrapin Bright Green
Browning, formerly from Rocky Mountain Institute, is a founding partner of Terrapin Bright Green, a company that best describes itself:

> Terrapin was created out of the Partners' shared sense of urgency to transition to a sustainable development model that could only be achieved by working with developers, communities and companies around the world. Since 2007, this concern has become an imperative to not only create a sustainable world but one that is aligned with natural processes and supports human health and well-being at all levels. Through our collaborative charrette process, we

Fig. 3.10 Judith Heerwagen (*Source* Author)

identify the science-based goals and aspirations of a community, develop an implementation plan, and then provide guidelines and other tools to support the team's success.

(Terrapin Bright Green, 2014)

I had met both Browning and his colleague, Ryan, at the Biophilic Cities conference and discovered that Browning had attended the Rhode Island conference from which emerged the book *'Biophilic Design'*. He contributed to the book and has continued, through his work, to progress the concept of biophilic design. Browning studied in the fields of environmental design and real estate development. Since then he has been active on an impressive number of councils, boards and committees, advocating for sustainable design solutions and developing strategies for business, government and civil society.

According to his website bio, "In 1998 Browning was named one of five people 'Making a Difference' by Buildings magazine. In 2001 he was selected as an honorary member of the American Institute of Architects, and in 2004 he was honoured with the U.S. Green Building Council's Leadership Award" (Terrapin Bright Green, 2014).

I was intrigued as to how he had come to be part of the Rhode Island meeting and a contributor to *Biophilic Design*. Browning was happy to share the story with me and it was to provide another insight into the progression of the biophilic design social movement.

Browning had founded Green Development Services at the Rocky Mountain Institute, an entrepreneurial, non-profit 'think and do tank' consulting on many notable projects. While at the institute he had researched worker productivity in green buildings, becoming interested in what conditions improved worker productivity. Browning had the opportunity to be involved in a study of the design of a large building for the Herman Miller Company. To improve worker productivity, William McDonough Architects focused on daylighting and providing views of the landscape for the workers. With research funding available, he saw an opportunity to measure any gains in worker productivity. A research hypothesis was created for the project and it was this hypothesis that was linked to biophilia. Three other researchers formed a team, one of whom was Judith Heerwagen. According to Browning, Heerwagen was responsible for a lot of the early biophilia research, particularly the savannah hypothesis.

The project utilised an existing worker population, with known productivity, who were being moved from the old factory to the newly designed building which had a focus on daylight and extensive views of rolling prairie, grasses and water features. Initial productivity results were 'messy', but it was when the worker shifts were separated out in the results that the power of biophilia was seen. With the night-time shift, there was no gain in productivity, while the daytime shift, who enjoyed the benefit of seeing outside, had significant gains. The swing shift had mixed results that were found to be seasonal. In summer, productivity was increased.

Browning has followed by a series of similar research projects, two of which were influential to him. The California Energy Commission employed a researcher to study the effects of daylighting offices, retail outlets and schools. The results were perceived as very controversial showing highly significant positive numbers, to the extent that the Commission asked for a rerun. The results came back the same, but even stronger. The researcher realised though that something else was at play in the results other than daylight and conducted a sub-experiment which extended the research to also focus on a Gold LEED building. The building already had large windows and plenty of daylight but the workers could only look out the windows if they turned. At the cost of $1000 per desk, the desks were angled so workers could see out the window to the trees' movement from their peripheral vision. Productivity increased 5–6% to a $2,999 per station per year return. The only variable had been the position of the desk.

Browning considered these two pieces of research the strongest on biophilia and workplace design. Coupled with Ulrich's earlier 1980's research on increased healing rates in hospitals with patients' exposure to views of nature, Browning was motivated to start a database at the Rocky Mountain Institute on research supporting biophilia. A co-worker left the institute to study with Stephen Kellert at Yale, sharing the Rocky Mountain Institute database, which led to the beginning of a conversation between Kellert and Browning on their shared desire to progress the emerging research into design. Ultimately, this progressed to the Rhode Island conference and Browning's involvement with this and the subsequent book *Biophilic Design* (Kellert et al. 2008a, b). "The book is really the first shot across the bow about biophilic design." Later, in looking at the design categories that were compiled in the book, Browning, along with a colleague, realised that these categories were unwieldy for designers. "They stepped back and saw that there were three clear design categories under which fourteen patterns sat".

The story continues when a client of Terrapin Bright Green and their partners, architects Cook-Fox, were constructing very high-end residential living. The clients were looking at living spaces that were about human health and biophilic design and asked Browning to look at the science of biophilic design. This led to Browning writing *The Economics of Biophilia.* A team at Google got inspired by this report and asked for clear design patterns so Browning shared the 14 patterns of biophilia with them. Google uploaded these into their design criteria for Google projects worldwide. These are specific to Google. Browning, with colleague Catie Ryan, decided to develop a white paper that examines these patterns and the science behind them. They gave me a sneak peek at what was coming.

I commented on how much these patterns and the concept of biophilic design resonated with me and Browning then recounted a story about applying for funding for the Rhode Island conference. He explained:

> We had a whole bunch of foundations turn us down and the response literally was 'So you want us to fund research into the intuitively obvious?' 'Yes,' replied Browning, 'that is exactly what we want you to do.' So it took years to raise the funds. This is one of the reasons we did *The Economics of Biophilia* because we would talk to clients about it and the response we would get was 'that's very nice'. What we were trying to get them to understand was, yes that is very nice and there are huge economic implications to this.

He considers that the economic savings with hospitals is one of the major examples of the potential benefits of biophilic design applied.

Browning has had conversations with neuroscientists regarding human response to nature and the patterns of nature. He says that positive responses can be seen by looking at brain activity but that neuroscientists don't know why this occurs. One leader in health care and evidence-based design Browning has spoken to consider that clinical certainty is not needed for something that is known and that it certainly does not have negative impacts.

> The two strongest desires we have come across from office clients is the impact on stress relief and cognitive functioning. We are looking at reducing stress through biophilic design regardless whether the stress comes from the built environment or the job itself. We are also looking at improving cognitive functioning and general well-being. For us, in many cases, the whole point of getting into biophilic design was just asking this question of: ok how can we design spaces that really support people's well-being? Biophilic design is one way to really help that.

We discussed the questions of how much exposure to nature is needed to have an effect and how long would the effect last? Browning says that studies are revealing relatively short periods of only 5 min will have an effect but we both agreed that this must vary with people and their attention. He talked about how visual connection with nature is important but the impact can vary with the depth of the view, and whether there is movement that would gain a person's attention, especially in the peripheral vision. Again this comes back to the evolutionary survival response of needing an acute peripheral vision to see possible dangers coming from the side. Browning talked about the pleasure side of viewing nature, citing Biederman's (2006) research which revealed that viewing certain patterns of nature fired our opioid receptors.

Fig. 3.11 Bill Browning on the COOKFOX green roof with beehive in New York (*Source* Author)

Browning thinks that while city parks are important, the focus needs to be on giving everybody access to a bit of nature every day and this can also occur in the little spaces. This is another of Browning's motivators: "Ok let's take it down to the scale of a commercial interior, let's take it down to the scale of the home, let's take it down to the scale of the house on the street".

Browning's colleagues installed a green roof outside their office space office space in New York 7 years before and "it was exotic but our neighbours started looking at it and now you walk out on our green roof and look around and you see green roofs all around. All over North America people are discovering their roofscapes and that they don't have to be asphalt and rock and mechanical" (Fig. 3.11).

We discussed research that had emerged regarding biophilic design and its influence on reducing rates of violence in both housing estates and prisons. Browning knew of some research carried out in a Portland hospital where pictures of nature and a window to the garden in one emergency waiting room had reduced both violence and stress. Browning's colleague had heard a story where they are putting pictures of trees on the security bins at a Chicago airport to reduce the stress of the whole security system:

> You need place based solutions, sometimes the intervention might not be something alive. Fractal patterns are extremely effective and perforated shade screens that mimic natural light patterns. These shade screens can also protect plants that are on the inside.

Browning had read some research on the relationship between colours and creativity. Green was found to be the colour which led to the highest creativity. He mentioned research on fractal patterns and that the ratio people most positively responded to is the ratio associated with plants and the skyline of plants. We continued discussing the designs of biophilic initiatives and which were most effective. This led into a brief conversation about beauty and the importance of aesthetics in nature and in biophilic design.

I asked Browning what he saw as the future of biophilic urbanism. "Now more people live in cities than rurally for the first time in the history of the species and within a couple of decades it will be seventy percent. So, for me, that makes this really, really important. The ability to get people out in nature is going to get harder. So it's how to get nature to everybody, not just the rich".

Browning grew up living in different locations around the world but most memorably in a small German village where he spent a lot of time playing in nature. He recognises that this has changed for many children in the generation since.

Browning is a groundbreaking researcher, fascinatingly knowledgeable with a passionate and inquiring mind. The conversation I had with Browning at their New York offices was one of the highlight meetings of my US journey. The learning I experienced in a few hours was exceptional, illustrating the knowledge and commitment that Browning has in this area. He introduced new concepts and deepened my thinking on others. When he mentioned the intuitive knowing he experiences and has witnessed with others, it resonated with me, and was the first time I had heard someone articulate this. I have heard so many people say that biophilic design makes sense that surely this reflects something felt at a deeper, intuitive level. This and other concepts and ideas that were mulling around in my head were brought to light and aired in this conversation.

Browning supports all his conversation with stories of research he has read or been involved with. He has researched through his own interest and in response to gaps in the literature or client need. Browning's creative thinking has been a significant part of the emergence of biophilic design but he also supports this by driving implementation through his company.

After the interview, he took me out onto the green roof outside his office. It had been built as a trial and showpiece and to provide the office workers with a green amenity, in line with Browning's design approach. They had noticed the number of bees attracted to the roof so had also installed a beehive. Browning pointed out the number of green roofs that the neighbouring buildings now also had installed. It was wonderful to see in the heart of Manhattan Island.

3.4.2 The City Visits

Washington DC

During the Biophilic Cities conference in Virginia, participants had the opportunity to visit the Anacostia Watershed project in Washington DC. Conference delegates canoed the river listening to the stories of the work of the Anacostia Watershed Society, a non-profit organisation working to conserve and protect the land and water of the Anacostia River system in the District of Columbia and the State of Maryland. Their efforts to clean up the polluted Anacostia River have resulted in unanticipated social benefits with the connecting of the diverse social sectors of the west and east banks. Social projects conducted by the Anacostia Watershed Society have helped raise the sense of belonging to the river and the city for the lower socio-economic east bank population. Riverside revitalisation initiatives reflect biophilic design and have included a small green roof (Fig. 3.12). An arrangement was made

Fig. 3.12 Beautiful and successful initiatives by the Anacostia Watershed Society (*Source* Author)

with Michael Lucy, senior consultant with the Anacostia Watershed Society, to meet for an interview when I returned to Washington DC after the conference.

Michael Lucy—04/11/2013
Senior Consultant
Greening the Built Environment
Anacostia Watershed Society

Upon my return to Washington DC, I was able to spend an afternoon in conversation with Michael Lucy of the Anacostia Watershed Foundation and some of his co-workers. He gave some background to the development of green roofs in Washington DC and the Foundation's role in the implementation of these. In 2005, the Anacostia

Watershed Society hosted a *CitiesAlive* conference. I asked why this occurred and Lucy explained "There were a lot of people interested; Parks and Recreation, the Mayor was interested in promoting these things and the District Department of the Environment. This was to encourage more green roofs as they make such a difference for storm water management plus the other environmental benefits, it just made sense". Although green roofs had been appearing in Washington DC for the last 20 years, the conference catalysed the interest and growth so that at the time of my visit there in 2013, there were over 2 million square feet of green roofs over 200 buildings. The US Coast Guard green roof itself is over 10 acres. As part of the Sustainable DC initiative, the plan is to have 20 million square feet of green roof by 2020. Lucy thought that this was a "huge stretch since it has taken at least 10 years to reach the current 2 million square feet". I asked Lucy if he considered valid the rippling effect that successful green roofs can have in catalysing further implementation. "Absolutely! People see one next door and think they can have it here, absolutely!" he replied. Lucy is hopeful that the Washington DC universities will install green roofs as well as more of the embassies.

Currently, the English and French embassies have green roofs and there is hope that the Chinese and the Australian will follow Lucy thought that the embassies want to "look cool as they are representing their countries and they want to have a green building. The European countries have known about it for a long time". The French embassy, Lucy explained, had undertaken a few sustainability initiatives and the green roof was considered "a simple initiative" for the benefits that came with it. "Of course the green roof rebate helped".

Lucy explained the development of this rebate. The Green Roof rebate program started in 2005 and was originally overseen by the Chesapeake Bay Foundation with funding from DC Water. The District Department of the Environment (DDE) started their own rebate program around 2007 and the Chesapeake Bay Foundation rebate closed down. For the past 4 years, the Anacostia Watershed Society has been managing the DDE rebate program for them, doing outreach and publicity initiatives. In these past 4 years, 30,000 square feet of green roof per year has been funded.

The DDE's main interest is "to manage storm water more effectively" and this is considered the primary driver for green roof initiatives in Washington DC. "The money and the intention is to get these roofs installed so that the water gets slowed down and absorbed and cleaned. And there are all the other benefits, clean air and good for the building, people are happier and healthier, properties are worth more", Lucy explains.

The conversation went back to the significance of the Green Roofs for Healthy Cities conference. The rebate came shortly after the conference and Lucy considers that "it helped" towards the rebate being implemented. The rebate is currently \$8 to \$10 per square foot and requires suitable plants that will do well, to be grown. Lucy says that they have talked about having extra funds available for native plants but "because green roofing is so new in America and there have been very few long term studies on native plants done, we have not required it. We want to get as many roofs up there as possible so we can manage as much water as possible. There is funding

Fig. 3.13 Green roof initiative towards storm water management and reducing water pollution (*Source* Author)

available for innovative proposals for greening the built environment if people want to experiment on green roofs and biodiversity" (Fig. 3.13).

I asked what Lucy saw as the barriers for green roof implementation and he had many thoughts about this. "Many people are completely separated from the environment so they forget how important it is and how connected we are to the natural world. Here in America we have spent many years designing systems and homes to get water far away from our buildings as quickly as possible. People are afraid of their homes becoming wet with leakage. This is a new technology; people don't understand it very well so they are afraid. So the tradition of building in a certain way means people are not used to this way of building. That is the biggest impediment I think and the second biggest is that it costs more. It may cost more up front but there are longer term benefits. The roof lasts two or three times longer. There is less heating and contracting and the roof actually can leak less. The benefits are not packaged very well".

Lucy had heard research that suggested that the biggest economic benefit of green roofs is "reduced temperatures, therefore reduced fatalities. Less people were dying. Even with the slightest reduction would have more people survive plus have reduced health requirements. On the hottest days our elders and young kids have the biggest health risks. Green roofs can help".

The interview with Lucy provided a practical understanding of biophilic design in action. Although the driver here was an environmental one and not 'biophilia', Lucy demonstrated awareness of the human–nature connection and the current disconnect from nature that has permeated design of our built environment. It was encouraging to see how financial rebates can aid the implementation of a green roof as a strategic choice in dealing with the cities' crisis of water pollution and storm water management.

Kelliann Whitley—DC—05/11/2013
Senior Property Manager
Blake Real Estate

Lucy, the previous interviewee, had provided names and locations of some of the green roofs in Washington DC which were possible to visit and, in viewing one of them, I unexpectedly met a remarkable woman. She worked for a real estate company located in a commercial office building. The woman, Kelliann Whitley, had read about the green roof rebate program and "saw the sense in it". Whitley thought a green roof would be good on the commercial office building where she worked in downtown Washington DC. Her preliminary efforts to have one installed initially met a lot of resistance. First Whitley had to convince the owners of the building. It took her a year of pushing through obstacles and continual research. "The owners of the building were afraid of leaks and voiding the warranty of the roof membrane so I rang all around the world, even Mayor Daley's office in Chicago to find out how to do this". This resulted in the roof getting a new membrane. Whitley was then able to secure funding to enable six young teens from abusive backgrounds to be trained to plant the roof with sedums. A condition of this funding was that the roof be open to the public for 2 years as it was considered a demonstration project. In 2015, it has been open for 10 years and Whitley still conducts tours, having shown roughly 6,000 people around the rooftop. She says "the owners of the building now 'brag' about their green roof".

Whitley has observed a steady increase in caterpillars, dragonflies, bees and birds on the rooftop. "It attracts and supports wildlife on K Street". She considers that "since it first opened its success has encouraged developers and government agencies to take the risk on installing a green roof. Now the process is much easier".

Whitley's passion is tangible. Through her journey of the installation of the green roof, she feels she has "rediscovered herself and found a love of the environment that I can't get enough of". On Whitley's roof tours she advises young adults to study green roofs as she views it as the future of the planet. She, herself, is now certified in green roof maintenance (Fig. 3.14).

> Ten years later and I am still really in love with the roof. When I leave the office, I sit by the roof and look and feel in a different world. Like I am in heaven. Habitat and beauty on top of an ugly building, once you learn about it, it is addictive. People want to see a living thing on top of a building. And all it does for the environment is positive. Mentally and physically it is healthy, peaceful and unique. Really it is a no-brainer.

Whitley is unique in these interviews as she is not affiliated with any university, organisation or industry involved with sustainability or design. Her driving passion and persistence in making the green roof happen were very inspiring. Whitley demonstrated qualities through which a person can quietly become a local champion. Her expression of the feeling she felt when on the roof and how important the connection to nature she had discovered was validated further what biophilia and biophilic design represent.

Pennsylvania
Helena van Vliet—07/11/2013
Biophilic Architect

Van Vliet is a successful Pennsylvania-based architect whose personal journey and intuitive knowing have guided her architectural designs towards a biophilic focus.

Fig. 3.14 Kelliann Whitley and her green roof (*Source* Author)

"In my work I was always trying to connect people with nature. I was always trying to build buildings that were in natural materials, rooted to the ground and would draw people outside no matter what the weather, which breathed. I was also interested in the connection between buildings and health". She grew up very connected with nature and drew on these childhood experiences for inspiration.

Van Vliet came across the term biophilia 6 years ago while searching for a term to describe her approach to her architecture. "I was redoing my web site and people were asking what style I do. I said I don't do a style, it's more of a process. But what is it that I do? Then I discovered this word and I read about it and thought yep!, that's what it is. I started to read about it more and more".

In the last few years, van Vliet has become increasingly interested in the health and well-being aspects and really connecting architecture with health. "Architects are not unlike doctors and really have a tremendous influence on the sense of wellbeing of occupants of the spaces and buildings that we create". She considers architects have a responsibility to design buildings that support health rather than increase stress. She now lectures on biophilic design in both architecture and medical professions. Van Vliet tells her architecture students, "What you will be doing will make people either well or stressed so how do we design places that are nourishing or relaxing. As Kellert says in his books we will only sustain over time buildings which we love".

Her design focus is on enabling people to connect with nature through their senses, not their intellect. "Places where people feel at home in and feel safe in and feel well in, invigorated and playful". I asked her whether her designs were driven by intuition or by the literature. "It's both, it's the intuition that drives it and then I go and find the literature and I think yep! I try and find the connection between the client's needs and the site's needs. It is inherently there, I just need to find it".

At times, van Vliet does encounter some resistance. "Sometimes the way I design is more expensive because it involves organic shapes and curves. Average builders can be frustrated because they haven't built this way, or I have roof forms which are very complex. The cheapest building is a box with a lid on it, so you add cost when you deviate from that and you have to have a client who understands that". Van Vliet knows of other architects who work like her, "but they might not call it biophilic design".

The concept of what we perceive as beauty is a large part of van Vliet's personal research. She cites Wilson in his definition of beauty. "Beauty is our word for what had been most successful in survival. If something is not beautiful, it is probably not efficient in a wholistic sense. Beauty is not subjective, it is part of our evolutionary design. A lot of our illnesses are stress related today. We can create spaces that are nourishing and helpful instead of depleting". Living in a small rural Pennsylvania town she wonders how we can house animals in nature zoos yet we still house ourselves "in concrete jungles where people walk in circles".

She advocates for biophilic design through her blog, website and by hosting discussion groups in her exemplary, self-designed home (Fig. 3.15).

Talking with van Vliet was like immersing myself in the beauties of nature. Her depth of knowledge and understanding about the human–nature connection and how to encapsulate this in design was captivating. The intuitive knowing and understanding was a strong feature of the conversation as I think there was a shared acknowledgment and awareness of this connection. We both have a wonder and deep bond with nature. It was inspiring to hear her journey and her work in expressing this passion, and of her joy in discovering the framework of biophilic design. Although her designs were always nature based, van Vliet now has a term for her work that she can share. Her explanation of the importance of aesthetics and of the theories connecting our evolution and beauty was illuminating and thought-provoking.

New York

Mary Rowe—15/11/2013

Managing Director and Head
Global Initiative for Urban Liveability and Resilience
The Municipal Art Society of New York

I had contacted Mary Rowe about a possible interview after reading a blog she had posted titled 'Urbanophilia and the End of Misanthropy: Cities Are Nature' (2013). Her article had prompted some deep reflection, presenting an interesting perspective.

She argues for self-organisation, supporting Jane Jacobs' theories that cities are dynamic, living things. They are created by the people who live in them in a way to sustain and nurture them and this is called liveability. Rowe suggests that perhaps sometimes policy can get in the way of people's innovation and resourcefulness. Good policies, though, should support these natural processes. But this made me

Fig. 3.15 Helena van Vliet in her beautiful home and admiring a green wall at Longwood Gardens (*Source* Author)

think about the mechanistic, sterile hard-edged cities removed from nature which humans have constructed and why these have occurred? Have the city inhabitants been removed from nature themselves to such an extent that this has led to this approach to city design? Or has it been driven by economics? Is biophilic design being embraced so rapidly as part of self-organising and an innate desire to not only reconnect with nature but to save our cities from the crises that some are facing? Rowe suggests that small innovations come from productively adapting to challenges and this not only increases liveability but also increases resilience.

Rowe began our conversation by telling me about a New York resident who has mapped the vacant lots of New York and seen that they perform an ecological function in providing biological corridors. This introduced her thoughts about self-organisation and technology, a topic she is obviously passionate about. "Digital devices have enabled whole new approaches and there is a meshing going on between the sharing economy and people developing 'apps' and looking at digital ways to manifest those kind of connections. So that, in my view, strengthens the organic nature of the city because connection is becoming highly digitised as well as being spatial. It is a merging of worlds".

We talked about the 'bottom-up' solutions with Rowe expressing how this is where she sees the solutions coming, not from high-scale technological innovations that she reads about in resilience theory literature. "We can't wait for these large scale investments, we need them, but the only way to get compliance in those and while we are waiting for them, is that we need to mobilise people on the ground to do smart things and be really resourceful". Rowe calls these 'hyper-local' approaches which are about building social capital. Digital technologies are making this possible. "The ubiquity of technology is going to quite naturally enhance the speed at which cities become much more economical as far as energy consumption and in engaging people's ingenuity and actually seeing the city differently. This is the good thing about biophilia, it's a way of looking and a way of identifying assets in the city and finding ways to preserve and cultivate them".

Again the conversation returned to what Rowe saw as the value of technology in uniting cities and biophilic design. "Technology is enabling us to see more. I don't think it is taking us away from nature, I think it is connecting us with it". I mentioned that in my experience I perceived social technology as disconnecting people from their surroundings. Rowe agrees, so she clarified the type of use of technology she was talking about.

We discussed the creation of the many community gardens in New York. "They are occurring for all of it, for food, connection to nature, community, all of it". Rowe told me about the growing number and different uses of these garden spaces. I asked her what she thought was driving the increase and she related a program she had seen on the sharing economy which suggested it was driven by scarcity and need. Rowe considers it is more than that, though. She talked about the spatial constraint of Manhattan being an island so she wonders if there is a "heightened recognition that there is an interdependency. We may be doing different things but we want to work in the same physical space. So there is a kind of meshing. So we want to garden together". Both of us were intrigued by the understanding that scarcity, need, technology and interdependency were facilitating a change in the way we occupy our cities. "There is something all happening at once; there is kind of a confluence". Somewhere biophilic design fits in here as well.

Rowe considers that one shift that has taken place is the recognition that the answers are urban. "The answers are to be found in the city". We talked about the ideas of beauty in the city and organised complexity. "Landscape and topography can not only be beautiful, but also highly functional. People can want both". Enabling policies can promote the function as well as the community, but where do they meet? "Bottom-up and top-down tends to meet around technology, around data. Data needs to be made available to the community so they can analyse it and put it to good use. It's about building urban literacy. Policies need to be transparent for community".

Rowe gave examples of solutions and innovations that have occurred where local communities have been given the information to enable their contribution. "These are hybrid solutions. It's merging creativity and expertise into social capital". Rowe sees this as all about the culture of the city and is active in networking with global cities. She advocates for urban innovation to be cultured locally, adaptive and supported by public policy and investment.

The conversation with Rowe was fast-paced and presented ideas from a different angle to what I had experienced. Her ideas around technology were interesting and we were both advocates for allowing and supporting social innovation through policy and information. A shared faith in the 'wisdom of the masses'. Rowe mentioned creativity as a component of social capital, and it is here that biophilia sits: connecting people with nature fosters creativity.

The New York High Line

Early in my research, I had discovered biophilic projects that appeared to be particularly significant with reverberating impact, sometimes globally. The New York High Line was one of these projects, with a fascinating story of its creation. Although I could not arrange to meet with the co-founders of the High Line, I walked it three times and talked with staff and other visitors. I also purchased the High Line field guide which details the history and creative stages of the walkway. This is a synopsis of the guide.

The High Line was built in the early 1930s as a means of improving both congestion and safety on Manhattan's West Side. It elevated the railway freight line ensuring the continuity of being able to bring meat, milk and baking supplies into the city. It was nicknamed the "Lifeline of New York". Forty years later, freight rail began to decline and the last train used the line in 1980. The High Line then sat unused until gaining attention in the 1990s when property developers expressed interest in its demolition. By this time, the High Line was a self-seeded landscape, with seeds brought by birds and blown from the nearby Hudson River.

At a community hearing regarding the future of the High Line, local residents, Robert Hammond and Joshua David, discovered that there was nobody attending who was advocating for the retaining of the structure, so they quickly founded '*Friends of the High Line*' with the aim of preserving it. Able to rally other supporters, the

Fig. 3.16 The New York High Line (*Source* Author)

next 12 years Hammond and David worked with New York's Mayor and the New York City Council to successfully preserve the High Line.

The first section opened in 2009, with the second in 2011 and a third section in 2014. It is nearly two and a half kilometres long, with plantings reflecting the self-seeded landscape that grew when the trains stopped running. More than 300 varieties of perennials, shrubs, trees and wildflowers attract bees, butterflies, birds and other insects. In 2012, the High Line welcomed more than 4.4 million visitors (Fig. 3.16) (Friends of the High Line 2013).

I had the opportunity to walk the High Line in both autumn and early winter. Although numbers were less in the winter there was still a steady stream of walkers and activities. In conversations with fellow visitors, their appreciation of the High Line was apparent, especially with New Yorkers. One visitor did mention that surrounding rents had gone up for tenants, signalling to me evidence of people's willingness to pay for views of nature and/or proximity. Gentrification of the area had upset some long-term residents who could no longer afford the rent (Fig. 3.17).

The staff loved working there and all who I spoke to were very happy with their job. The number of visitors was definitely growing (Fig. 3.18).

This was a project that may not have come to fruition if two local champions had not attended a community meeting and galvanised community support for the idea.

Paley Park—New York

Browning had advised me to visit a small pocket park in New York located on 53rd Street in Manhattan. I visited 2 days after our conversation. As soon as I entered I was struck by the immediate transition from bustling streets to a place of peace, similar

Fig. 3.17 Autumn on the High Line (*Source* Author)

Fig. 3.18 High Line staff member (*Source* Author)

to stepping into a church. I was also struck by the biophilic features, many of which Browning had mentioned. The park has a waterfall, green ivy-covered walls and a 'savannah effect' canopy of honey locust trees. The sight of the water was enhanced by the white noise of the water which blocked the street noises. Looking at the people quietly sitting there, it seemed apparent that most felt that same sense of peace and refuge. I was able to talk with the resident caretaker about people's responses and uses of the park, and my instincts were confirmed. The caretaker mentioned how many people "felt it was a place of refuge, where they could escape the city and sit peacefully. People come for lunch. It is an oasis from work where people relax and clear their mind so are in a better frame of mind when they return back to work. The sound draws people in. They also come for romance". He said meetings and interviews were also held here. "It feels safe" (Fig. 3.19).

New York is a unique city. I had only two face-to-face interviews but both gave me a taste of what people relate to when they talk about the city: the strong sense of community that is present in the city. It is in this civic, community arena that initiatives are occurring such as the High Line and people choosing to green their own roof. It appears that in New York there is a high value on access or views of nature for the wealthier, whereas scarcity and need for food drive the less wealthy to create community gardens, connecting with nature in this manner.

Fig. 3.19 Paley Park New York (*Source* Author)

Chicago

Arriving in Chicago held a sense of excitement. It was home of the iconic green roof on Chicago City Hall and I had organised an interview and viewing of the roof. This roof I had discussed in presentations previously, as the drivers and process for implementation, was a story in itself which conveyed knowledge and inspiration. Plus it had needed local champions.

Chicago itself is a challenging city as the windy climate ranges from cold and snow to life-threatening heat.

Jay Womack—26/11/2013

Director, Landscape and Ecological Design
WRD Environmental, Inc.
The Chicago Center for Green Technology

Jay Womack is a Director of the company WRD Environmental with ecological landscape design as his area. WRD Environmental is an ecological consulting firm that creates and fosters environmentally responsible urban landscapes. They design, install, manage and consult on projects and programs that conserve natural resources, promote sustainability, cultivate biodiversity and restore nature's balance.

Womack is also an Adjunct Professor at the Illinois Institute of Technology where he teaches in Landscape Architecture.

I asked him if he knew the word biophilia and was wonderfully surprised at his response. Not only was he aware of biophilic design but he taught it as a unit in his landscape architecture courses, espousing E. O. Wilson as one of the great thinkers of recent years who he is inspired by. Aldo Leopold's *Sand County Almanac* also had a big impact on Womack's thinking. It was through these readings and self-education that Womack has developed a deep understanding of biophilic design, "These concepts resonated with me". He had translated these understandings into course outlines for the landscape architecture students, as well as sustainable designs for clients.

His description of what he sought to convey to his students best describes his perceptions which have motivated him to advocate for biophilic urbanism:

> When I start teaching the students I tell them right off the bat that I am going to take you on a ride that you are not going to believe, this is not going to be a course about landscape architecture, this is a course about ecology, ecological thinking, sustainable design and philosophy. By the time the course is over, they get it. Then I take them and show them projects I have worked on personally and how design is married with ecology, with storm water management, with creative structure. You can't just teach that out of a text book, you have to go see it, you have to smell it, you have to touch it, you have to feel it. You have to feel it in your heart; you have to feel it with your fingertips. I believe that a lot of our responsibility is to the earth and to the people and sometimes that does take a person who is more in touch with their heart, than with a pencil.

Strong in Womack's ethic is the message relayed by Aldo Leopold: "That a good landscape steward is not only someone who heals the earth but who also teaches others about how to heal and who shares that message. This is a huge part of what I stand for".

Womack considers that most landscape designers would not have heard of biophilic design; that landscape architecture does not teach ecology, it teaches design. "When the line from design into ecology is crossed, there is a whole different take on landscape architecture". He considers that for teaching professors to go back and think about a whole new design process can be a little daunting. Womack, though, through the nature of his being and personal interests has found landscape architectural design works best "if it goes beyond design and includes ecology".

While Womack considers Wilson defined biophilia admirably, he thinks people have "different interpretations" so has to consider what a client would understand. He finds that "regenerative design" is a term that people can relate to. For Womack, sustainability is about using less of something but he argues that now we need to raise the bar and go above this to designs that give back. "We cannot continue to keep using small sustainable amounts, we have to utilise regenerative designs that begin to put something back". This has led Womack to landscape design utilising native landscape concepts and plants, rather than ornamental, as he has found they contribute to the landscape. "They rebuild the soil and biodiversity, they help to absorb water and sequester carbon dioxide".

Doing this as a landscape designer had a challenging beginning. Womack had struggled with people's perception of natives "as weedy, and indeed, if not treated correctly they can be like this. Typically people wanted neat, clear, crisp organised gardens". One major client took a chance on Womack's ideas and allowed him to plant natives in a previously very large turfed area. "After about six months employees were going out into the area, eating lunch out there and coming to work happy. This is biophilic design". The project was viewed as a successful example.

He considers that "you need to get people excited about change, help them to approach it differently, before you can expect them to change". In Chicago, Womack thinks biophilic design is "borne out by architects because so much of biophilic design can be developed in a building's skin or an envelope or an interior. Whereas with landscape architecture it is a little bit more difficult as we are still being thought of as the step child of architecture". He does recognise that slowly there is an evolution of landscape architecture to be more important in developments. Developers are beginning to understand that "landscape architecture can assist in on-site water management".

Womack recounts the story of Mayor Daley's visit to Germany and how on his return, inspired by the German examples, he was determined to construct a green roof in Chicago. "This was in response to urban heat, but also for storm water management. Green roofs are now starting to be known a lot for air borne particles, air borne pollution that the plants will absorb, pick up and retain. I think the longevity is starting to become the life cycle analysis. They are being seen as a life cycle cost that is beneficial to a long term owner".

From the initial green roof that Mayor Daley instigated, Womack has witnessed the growth and momentum of the 'green' (as in nature and biophilic) movement. He considers that there "is now peer group pressure to be green and to care for the environment, that part of the motivation is not to be left behind. Clients now expect green. Great green projects have been getting acclaim, doors have opened for green projects and there are financial incentives".

Green and biophilic infrastructure has now proved to be successful in Chicago, not just Germany and Europe. "We have had to prove that things aren't just done in Europe. That they are happening here in the US and they work for the long haul. This is being recognised that these aren't bad ideas, they work. So it's now ok to write ordinances to enforce green initiatives".

Womack had a rural upbringing and has always loved to be outside. He says he "has always had the environment in his heart" and volunteers as a land steward.

The depth of Womack's connection and love of nature was apparent and I really enjoyed talking with him. He is a deep thinker and observer of nature and I admired how he had translated his understandings into a framework that he could communicate with others. Indeed he felt very motivated to teach and share his insights at the same time being very aware of the challenges in potentially changing people's approach. When he recounted how he had needed to change people's perceptions from seeing natives as weedy, I thought how again aesthetics had risen as an issue. Yet here it was an issue where people thought a certain look was beautiful and another not. Womack had successfully challenged this and provided a landscape where people not only found beauty but also connection.

In my conversation with Womack, he mentioned the visionary former Chicago mayor, Mayor Daley. Daley had a plan for a green belt through Chicago that was slowly taking shape but is renowned for his implementation of a green roof on Chicago City Hall. It was after seeing green roofs on a visit to Germany that he was inspired to bring them to Chicago primarily as a way of mitigating urban heat island effect prompted by a severe heatwave in which 700 lives were lost.

Although Mayor Daley had left office, I was fortunate to have an interview with the man who worked closely with him on the project, Michael Berkshire. Berkshire works in the Sustainable Development division for the City of Chicago. I had read a transcript of a presentation he had given about green roof policy following on from the construction of the Chicago City Hall green roof. So I felt I had a bit of background.

Michael Berkshire—26/11/2013
Sustainable Development Division
City of Chicago

Berkshire continued the story of the Chicago City Hall's green roof, corroborating the fact that it was "in response to a heat wave we had and the associated brown out. I think over 700 people died. So with the Federal EPA, we were looking at ways to lower urban temperatures by green roofs. In a densely developed city it is one of the few ways that you can do this". This was supported by Mayor Daley's trip to Germany where he learnt extensively about green roofs. "We needed to see if it would work in our climate and our political environment". In 2000, the green roof was installed on Chicago City Hall. "It was a huge success. It's been very well received and very well covered by the media. I am just amazed at the people that visit here and know about it. It has become our icon for our sustainability movement" (Fig. 3.20).

Fig. 3.20 Michael Berkshire and Chicago City Hall green roof (in winter so more brown) (*Source* Author)

In the early 2000s, the City was negotiating with private developers to incorporate green roofs as much as possible. Mayor Daley became "frustrated with the lack of progress", so employed Berkshire to write a policy to formalise the initiative. Since June 2004 they have had the sustainable development policy which means that if someone is receiving any type of assistance from the City, whether zoning or financial, all are required to contain sustainable elements in their project. The Mayor "was adamant that if we were going to provide them with something then they needed to provide us with something". The requirement began with green roofs and has expanded to include LEED certification.

Chicago City wanted to investigate the success of their policy initiatives so in August 2010 employed satellite imagery which showed that Chicago had 359 green roofs totalling five and half million square feet. They also wanted to investigate why those green roofs were built:

> Were they built because of the policy? Were they built because it was voluntary? Was it because of the storm water ordinance? Or was it because of our green permitting program? What we found out was that 82% of those roofs were due to the sustainable development policy. So that has been our main driver. We are assuming that many of our LEED buildings are because of the policies as well. We have found that regulation has been more effective than incentive. The stick rather than the carrot.

Berkshire concludes this may be due to the fact that they were early adopters of green roof technology. He remembers that in 2004, when they were holding focus groups to discuss green roofs, "no one knew what a green roof was and there were a lot of misconceptions. It was seen as risky to the development community". This is why Berkshire thinks regulation worked better than incentive.

They use TIF (Tax Incremental Financing) financing a lot. The central loop in downtown Chicago was covered by this and here Chicago City offered existing buildings a 50% match for installation costs for a green roof. This incentive only had one taker.

Another existing incentive is the green permitting program. If a building qualifies as a green building then the building permit may be reviewed and approved in 30 days or less. A green roof helps qualify towards a green building. Another successful ordinance that has been in play since 2008 is their storm water ordinance. Every project over a certain size has to submit a storm water plan that proves that they are either keeping a half-inch of rain on site or they are increasing the permeability of the site by 15%. Again a green roof helps achieve this. There is also a density bonus which means that a developer may build a bigger building if they put a green roof on it.

Berkshire mentioned another 'tool' called the 'Open Space' requirement for new residential units. They either have to provide new open space or they have to pay an open space impact fee. If they build a green roof that is accessible to residents, they get credit on that fee. He explains the success of this program towards green roof installation as developers see that they get credit plus they see the green roof as a nice amenity for their residents.

I asked him what response there was to all these policies from developers and he recounts early times:

> One developer when he came in was literally spitting mad. Literally spitting at me. Probably 3 to 4 years after the policy was implemented one of the most well respected zoning attorneys saw me one day and said. "You know before every time I saw you my blood pressure rose because I knew you would be asking for all this stuff that my clients wouldn't want to do. Now I am always very excited to present to you, as now, it's become a way of doing business in Chicago.

Berkshire explains that this has extended beyond developers to customers who inquire about what green features a building incorporates. "It is no longer 'what the hell is a green roof', it is 'does your building have a green roof?' or 'Are you going to be LEED certified?' Sustainable, green policies are a way of doing business in Chicago". Berkshire considers that once you get people asking for these things, then the development community starts paying attention.

I asked Berkshire why he thought people were asking for these things and he responded with an anecdotal story he thought demonstrated people's attitude and illustrated their desire for "something beyond the label of green".

The Magnificent Mile Story

At 900 North Michigan Avenue, prime real estate in Chicago on the Magnificent Mile, there is a mixed-use building with both commercial and residential. A few years ago, they wanted to convert some office space to residential. The policy required the developers to provide open space or pay the open space fee. The developers decided to build a green roof garden that was accessible to all residents as a way of providing open space rather than pay the fee. They did this on the setback car park roof that was on the opposite side to Lake Michigan. The building is surrounded by other high rises and the developers received so much positive feedback that they voluntarily vegetated another larger car park roof nearby. The developers then discovered that the spaces, particularly the office spaces, were renting faster on the green roof view side than the lake view side. Ultimately, they vegetated the setback roof on the lakeside as well so that 100% of the rooftop is vegetated. This had started with the developers taking advantage of an incentive to provide open space rather than pay a fee, then receiving positive feedback followed by economic benefits.

Where there is research and data for economic benefits in installing a green roof, Berkshire will use them. He cites the faster healing rates in hospitals as one he has explained to new health care facility developers. There is recognition that this is helpful in supporting the policies and incentives. Interestingly, he thinks "there should be more stories like the Magnificent Mile as he finds the economic argument harder to make because payback time can be slow".

Millennium Park was an unattractive open parking lot and train station where commuter lines ended. A roof was built over this and a 25-acre park has been established on the rooftop. Berkshire cites from memory a study that measured financial benefit from tourism and the huge amount of real estate development, especially residential, that has occurred surrounding the park. He thinks "the gain was 3–5 billion dollars of economic benefit. Also the Plaza around the bean sculpture is now where everybody goes" (Fig. 3.21).

Every summer there is a tour open to anyone called Great Places, Great Spaces. On the tour, people are able to get access to spaces that are normally not accessible. One of these spaces is Chicago City Hall's green roof and the tour is always fully booked.

Berkshire considers that to be globally competitive, the city has to be liveable and that frequently liveability is being equated with sustainability. He finds visitors are often surprised at how beautiful Chicago is, how green it is, with all the trees and

Fig. 3.21 Millenium Park (*Source* Author)

all the planted median strips. Berkshire says they are currently revisiting their 1990s goals for the amount of open space they wanted to achieve in Chicago. With the increased density they are no longer trying to retain the goal of 2 acres open space per 1,000 residents in all areas of the city so the new framework they are looking at is providing connectors between existing open spaces. These would be pedestrian and bike-friendly, incorporating storm water management as well. Berkshire thinks "people will choose a path that has more flowers or trees. I don't think they are consciously doing it, it's just something that I think most people are drawn to. People just get an enjoyment that they are not even thinking about. They just innately, or 'biophilically' choose it. Though in summer it may offer more shade so just be more pleasant".

Berkshire then moved to talking about the Chicago City Hall green roof and what it meant to him:

> For me, to be able to go on the roof top… it is quiet, or quieter, and it changes with every season and I love going up there in different seasons to find out what's blooming. I come from a rural background but consider myself a very urban person and love cities. But there is still this calming effect.

Even in cities he acknowledges that he needs access to some form of nature as he enjoys the calming effect nature has. This progressed the conversation to the nature versus nurture debate and whether, once people have experienced significant nature, they feel the necessity of continuing access and exposure. For Berkshire, air quality is important and so he chooses to live by the lake. He explained that he had noticed cultural differences in people's need for nature:

> In some particular cultural enclaves the residents tended to pave front and back yards and not have gardens. Some neighbourhoods, particularly one Hispanic area, has little open space and at a public meeting for the residents which I had attended, the recognition was there of the health issue, particularly for children, of the lack of outdoor areas.

Berkshire noted that Mayor Daley was born on Arbour Day with a strong connection with trees and during his time in office he had 75 miles of median strips planted with over half a million trees. One of the things Berkshire admired was Daley's commitment to "not ever asking the private sector to do something that the public sector was not willing to do". Daley needed to ensure that green roofs were viable in Chicago so not only was the roof of City Hall vegetated but a dozen other public building roofs were also. Different green roof technologies were used as part of the trials. Berkshire thinks this was a very 'smart move' and he often took concerned developers to the City Hall rooftop to show them it worked.

It is obvious that Chicago's biophilic initiatives are very policy driven. Berkshire mentioned how Portland, on the west coast, has been driven by grass roots and incentives and is only now looking at policy. He recognises that they have been very successful though without policy and have more square footage of green roofs than Chicago. Chicago has more green roofs by number. Berkshire also perceives Chicago as being different because of Mayor Daley. "In regard to implementing green roofs, Daley said 'I want this to happen, so figure out how to do it'. In comparison to Toronto, Chicago did very little research; they just 'did it'".

When I arrived for my interview with Berkshire at Chicago City Hall, the first thing we did was visit the green roof. This was quite an emotional moment for me. I had heard so much about this roof and it was hard to believe I was actually there, standing on it. It was snowy and did indeed look like an ill-kempt weed lot. My own judgement of aesthetics caught me by surprise. It was the beginning of winter so of course the plants looked dead.

Berkshire was a pleasure to listen to with his clear narrations of the history of policy in Chicago, the challenges and the successes. His recounting of the attitude he faced in first implementing green roof policy led me to admire his fortitude and conviction. He was a champion, or was he a hero? If champions face adversity do they then become heroes?

It was wonderful to hear of all the ideas and policy initiatives that Chicago had implemented; the tool kit as Berkshire called it. But what made it more special was to hear how well it was going, how accepted and now expected these biophilic initiatives had become. The Michigan Mile story stuck in my head. So much in this interview demonstrated the research I had been reading. It was inspiring and encouraging and it appeared due to a few good leaders stepping into the social movement and giving it a good push.

Toronto

Toronto was one of the cities around which I already developed a sense that biophilic initiatives had progressed well in their urban planning and design. I was interested to discover why this had happened for this city and in what areas the drive had come from. Here I had organised to meet with many people and it proved to be a busy but informative visit. I found a consistency in the story of the emergence of green roof technology in Toronto that involved a core group of people.

Brad Bass—21/11/2013

Great Lakes Issue Management and Reporting Section
Environment Canada

Bass is one of the group of individuals who were central in the emergence of biophilic design in Toronto. He sees that advancement of biophilic initiatives in cities occurs because "you have a bit of a number of factors that come together, often through a motivating need. One is that there are a few individuals who have an idea and are going to build it". In Portland, it was Tom Lipton who constructed the first green roof and in Toronto it was Wolfgang Amelung who built the first interior 'breathing wall'. Bass called these people the "early experimenters. Following this, you have the raising awareness stage. This is what Steven Peck and I did".

> In many cities you have an issue and it turns out that a biophilic approach to design can address the issue in some way. So you have a driver that pushes a change in the way we do things. Heat was a big part of the driver for Toronto, but so was water. So you have this driver, you have the awareness raising and you have the innovators. Also there may be a research community that supports the needs of the local policy. The answers are already there when the policy people need them. The policy framework sets a context for more people looking. If it looks good it will attract more people. So there is a slow but steady growth.

In the early 90s in Toronto, researcher Roger Hansell, who was interested in urban heat, had built window shades out of climbing bean plants. Bass and Hansell had been discussing urban thermal dynamics and when Bass saw this concept of 'vertical gardening' he was drawn in. "When I saw this I thought wow, that is the missing piece". He said that a lot of people thought they were crazy but then a student wanted to work with them on rooftop gardening, as it was then called, so they started to investigate the concept. A local architect also became involved in the rooftop gardening project, introducing another called Stephen Peck. Over dinner one night Bass and Peck decided that they should conduct an initial study and Peck was able to secure funding to do this. This was in 1998.

The research being done by Amelung, Hansell, Bass and Peck raised enough awareness that, by 1999/2000, a few architects were recommending green roofs to clients. Bass considers that this happened because "they were either responding to a client request or they wanted to do one and found the client would pay for it. But their interest was, it either grabbed you or didn't as an alternative solution to an issue". The uniqueness was attractive to both clients and architects. He also thought that people were drawn to the concept "as it could give access to greenery plus inner city living, and they were prepared to pay more for this amenity". Bass does consider that this can reflect peoples need for a 'green spot' to be a little calmer, breathe better air and improve their mood. Increased food security is also a possibility.

By 2001, Bass was being invited to lecture to architects and planners on green roofs. Two years after that, the invitations stopped and he found it was no longer necessary as the concept of green roof and green walls had taken off.

Apart from architects and planners liking the concept for its uniqueness, Bass explained that water management was another driver "especially for the ground-based stuff which is water quality not water quantity, as all watersheds feed into Lake Ontario. Green roofs were providing a low impact development alternative and a better way to deal with issues related to storm water runoff".

The shift to policy, Bass considers, usually "involves one champion who is pushing it. In Toronto it was through Peck's engaging of municipal politicians which eventually led to Peck tweaking the ear of one of the Deputy Mayors who became the champion". Bass then told me a story that involved an architect whom I had first heard about in New York, Margie Zeidler as a further example of local leaders.

A maintenance man at one of the buildings she owns decided he would slowly add plants by varying very basic methods to the rooftop as it was an empty space. The roof filled in a random but beautiful manner to become a luxurious roof garden. The roof made the newspaper and Margie saw the value it added to the building so began to add further greenery. She then bought a new building and added a green roof to it plus an indoor living wall. She is astute with her money but saw the economic sense with these initiatives (Fig. 3.22).

Bass acknowledges that these buildings feel different and attract more creative tenants, a benefit that can be added to the list of the full range of benefits. "In Toronto a benefit analysis shows that this is all worth doing".

Fig. 3.22 The building and initiatives of Margie Ziedler (*Source* Author)

It was a great start to the Toronto visit. Bass seemed very clear about how initiatives had emerged and the core group involved. Again it was a demonstration of the role for creative leaders and the impact they can have on the direction of a city.

Roger Hansell—21/11/2013
University President at Noble International University
Emeritus Professor, Department of Ecology and Evolutionary Biology
University of Toronto, Canada

Roger Hansell is a Professor Emeritus in the Department of Ecology and Evolution with the University of Toronto. He is also Executive Director of the Noble University for Environmental Peace in Toronto. Hansell spent his youth immersed in nature, eventually studying to become a professional biologist. Encounters with other thinkers have helped develop Hansell's deep and diverse thinking. He considers that biological systems self-organise for efficiency, to minimise gradients and this is how it should be; systems need to fit with the environment. Hansell also had worked at the Institute for Environmental Studies with the University of Toronto and it was here that he met Brad Bass, who I had interviewed previously. They discovered that "they had a mutual interest in how to protect cities from climate change; in other words, how to self-organise to minimise the huge gradients that are going to be put across us".

Hansell and another colleague, Diswhait Ganguly, were also discussing how there was a failure of stability within our cities functioning systems and thus a need to further research how systems worked, and how humans, nature and economics could self-organise to function in 'stability' together. This, Hansell suggested, "entailed environmental peace – peace within, peace between individuals, peace between groups, and peace between countries, sitting in stability alongside the human economy and the environmental economy". Hansell, alongside five Nobel laureates, established an institute, which had just become a university, to teach environmental peace.

> We ask each professor to show the students what happens when you change one of the fundamental assumptions of the area of what is being taught. That is when the breakthroughs happen, and it creates flexible minds. We need this to solve the problems of tomorrow.

Hansell's interest in biophilic initiatives comes from an environmental peace perspective. He cites Romeo and Juliet Act 2 Scene 1.

> 'The air is hot, the blood is hot.' Temperature affects human behaviour. Research shows that when temperatures increase so do the number of violent crimes and domestic violence. Green walls and roofs are some of the solutions that give environmental peace, that is, they reduce temperature.

He recalls aerial photos mapping heat in Ontario cities. "A shock was that the round circles of high heat were found to be school playing fields comprised of astro turf." Hansell also cites research done by a Toronto magistrate on arrest rates and temperature. There is a direct correlation, when temperature rises so do arrest rates. He is concerned that with climate change and increasing heat "violence will reap the rewards".

For Hansell, the history of the green walls and roofs in Toronto goes back to the same early group of researchers whom Bass and Siber had already mentioned. Some members focused on feasibility studies for the implementation of green roofs and walls in Toronto. Others, including Hansell, agreed they just need to begin physically trialling initiatives. He expanded on Bass's story of their first green roof trial. With the help of students, they constructed a makeshift green roof over part of the University's faculty roof and left the rest of the roof exposed. They measured temperature variations, finding the exposed roof to bake at 50 °C, while under the green canopy the temperature was 20°. This gave them further impetus. A technological breakthrough emerged from Germany in the form of a roof waterproof membrane, which significantly helped overcome the barrier of potential roof water damage.

Hansell considers the main motivator for bringing more green walls, green roofs and nature into our cities as "*common sense*". He referred to the African 10 million tree story (where an African woman galvanised people to plant trees across borders) as an example of this common sense. "Politically, it would almost be impossible to get countries to agree to undertake such an altruistic but necessary joint project but one woman saw this was common sense and made it happen" (Fig. 3.23).

Hansell was a significant elder and oozed years of reflection and deep thinking that led to a diverse and fascinating conversation. I was enthralled by him from the moment we sat down and the conversation ranged through topics I never would have

Fig. 3.23 Roger Hansell, a wise elder (*Source* Author)

thought related, but were somehow, as they gave the connection to the big picture, the bigger issues, and perspectives which went to the core of human nature.

My conversation with Mary Rowe in New York had also talked about self-organisation. Was this social movement also about humans and the environment self-organising in response to a crisis? Hansell discussed what had been touched on before but from a different angle. This was intriguing and triggered some deep thinking in me as I grappled to place the pieces together. There was this deeper concept that raised its head in varying degrees and ways in my conversations. It was to do with nature's impact on our creative thinking. Around nature we are more relaxed, less stressed and function better. This tends to lead to more creative, innovative, open-mindedness and flexible thinking. They flow from one to the other and through this thinking, as Hansell suggests, the innovative breakthroughs come and the potential to solve the problems of tomorrow. To create a university, 'The Noble University of Environmental Peace', to teach your ideas, is testament to the wise leader who is prepared 'step outside the box' to convey years of experience and thinking to help future generations.

Birgit Siber—22/11/2013
Principal
Diamond Schmitt Architects

Siber is a Toronto-based architect who was pioneering in indoor green walls. She has "always had a passion for plants and how they might be able to be integrated into buildings so I had one ear open for what is happening". The University of Guelph in Toronto was doing research on how plants can cleanse air in connection with programs for the International Space Agency. "What was interesting to me was that they were studying this for space travel in extreme environments and yet it was applicable to us in our day to day lives because people spend so much time indoors and away from plants. In Toronto it is a staggering 90–95% of their time away from plants".

There was a project in Toronto to merge Humber College and Guelph University. The venture was to be housed in a purpose-built large-scale building on the college's campus. Siber was invited to consult on the design. The thought was to bring something of Guelph to the college site. Siber went to visit the scientists researching the living walls at Guelph. The walls they were trialling were small and unimpressive. The scientists advised her to not even try to integrate this into her building. Contrarily, when Siber produced a visual rendering of a possible living wall, she found that the steering team for the project, even the most conservative, "were intrigued, even compelled by the idea". Siber and her team of architects were stunned by the resonance and universal appeal they encountered. She thinks the "notion of having plants brought indoors on a large scale resonates. I think there is something fundamental and primal about being attracted to them".

Siber and her team took an idea and expanded it, suggesting a living wall 30 feet wide and 50 feet high. She refused to make it smaller or to do a trial but wanted to "go for it". Although, at this stage, Siber knew the steering team had not given approval but only agreed to consider it, she was determined to stay ahead in any questions or issues they presented to her. This made the path smooth for consideration to transition to approval. Also, as the amalgamation of the university and college was new, it needed advertising, and the render of the potential living wall Siber had completed was used in the promotional materials. "The university was receiving many enquiries and interest as to where this wall was, which further invigorated the argument in support for building it. It received approval and $400,000 to construct it". Siber appreciates the risk they took, but she also appreciates "that I was in the right place at the right time. I just had to hold the image and things seemed to flow. It had an energy of its own".

The wall was installed in 2004 and has enjoyed a fantastic response. "Students gravitate to it, sitting around to work and talk" (Fig. 3.24).

> Siber's architectural company has found many of their clients are interested in the concept of the living walls, often visiting the one at the college. She observes: The clients will find ways to finance a wall, even when they were having to cut costs in every other way. They were very reluctant to give up the notion of this living wall as it not only provides a beautiful feature but actually cleans their air, provides some evaporative cooling. Actually in the large atriums it provides some acoustic attenuation and you have the white noise of the trickling water.

Siber was keen to explain the unique function of this living wall, and the others she has since built. The wall removes volatile organic compounds from the air utilising the technology undertaken by the scientists working on the space agency program at the University of Guelph. One of the scientists received a fellowship to commercialise his research and enable him to construct living 'biofilter' walls. The scientist, Dr. Alan Darlington, spent time on the living wall construction and how the wall could support the plants hydroponically in a vertical plane with air moving through the plant

Fig. 3.24 The Guelph-Humber indoor biofilter green wall (*Image courtesy* Nedlaw Living Walls)

root system. In a two-layered system, he was able to place the plants with their roots in a position so that the air could be pushed through them and the water into a cavity behind and up into the mechanical system for circulation throughout the building. There is no toxic waste as the volatile organic compounds are removed through the plants' biological processes. The hydroponic aspect facilitates the biofilter process far more than potted plants (Fig. 3.25).

She feels the walls work on so many levels, and is "fascinated by how many benefits this integrated approach to plants in a building can actually achieve". Since its installation another 14 biofilter living walls have been constructed in North America. The prime motivator for her clients varies but Siber clarifies:

Fig. 3.25 Corus Quay indoor biofilter green wall (*Source* Author)

> I definitely think having a large scale feature that says we are 'green' is up there; that we are definitely doing the right thing and trying to provide and endorse green technology. I think it is a good poster child as it were for green initiatives. There are so many sustainable features in a building that we don't see. The clients also want to be part of a very interesting way of bringing plants into a building that delivers many benefits. The ability to deliver better air quality enhances the ability of people to focus, be more productive and it probably encourages less absenteeism.

There is a huge problem in North America with indoor air quality and sick building syndrome and the indoor living walls help rectify that. The field studies of a lot of the larger indoor walls are indicating that 85% of volatile organic compounds are removed from the air so it is of the same quality as outdoor air.

Beyond the beauty and aesthetics of the living walls, Siber thinks "there is something more than beauty going on. I think it is just tied right to our cores, that we need that connection to plants. We have created such an artifice around ourselves that it is becoming more difficult to be in those kind of settings". She wonders if not being around nature so much enables people to recognise how much they need it.

Siber grew up in a cottage in the mountains where she lived outdoors except to eat and sleep.

In Siber, I had met another leader who saw an opportunity and made it happen. She also was very connected to nature and had a love of plants. Siber's living wall stories presented a depth of personal understanding of a primal resonance with nature which she witnessed with her clients and which I had also witnessed in my interviews. It was this that seemed to drive the compulsion both by the committee and the public to accept the first trial living wall at Guelph-Humber University. It was also interesting to hear of an initiative different from green roofs and the idea of indoor biofilter living walls fascinated me. It brought in another worthy benefit to the growing list.

Steven Peck—25/11/2013
Founder and President
Green Roofs for Healthy Cities

From the initial group that came together, there developed the 'Green Roofs for Healthy Cities' organisation. Steve Peck is the founder and director. The world that Steve Peck inhabits is the one focused on developing the market for green roofs and green walls. He sees that biophilic design is just one aspect, one piece of the puzzle, but not the only piece. Peck acknowledges that with green walls, in particular, biophilic benefits are probably the most predominant benefit. He suggests that living walls don't offer the same benefits as green roofs but they offer a lot of strong biophilic benefits as they are "right in your face and quite striking when you come up against them. They have a biophilic impact on people. Lots of green walls are built because they need to manage storm water or there are regulatory drivers. They create green spaces in apartments for people and can attract higher levels of rent or increase the value of the units for sale". Green roofs bring a wide range of environmental benefits, "more mundane stuff, such as storm water management, reducing the urban heat island effect and cutting back on energy consumption, sustaining the life expectancy of roof membranes". Biophilia is not a driving force for green roofs. It may become the driver as people such as Bill Browning "unravel the science and economics of it". I discussed how the environmental benefits that he discussed are ones that myself and others consider biophilic, as they are natural systems integrated into our urban buildings. Even if they are not initially built with a human–nature connection motive, they are often discovered and enjoyed by those with access.

Peck then jumped to suggesting that I had taken the term biophilic urban design and put it on top of a lot of things that weren't biophilic, that there is no such thing as biophilic urban design. I said I had met some biophilic architects and designers. "That is all just smoke and mirrors".

He suggests, though, that the emerging science of the economics of biophilic design is what will change how our cities are designed and it will help bring vegetation into our cities. It may evolve into a school of biophilic design. As far as a social movement of biophilic urban design "This is coming from Australia. You do things strangely down there, upside down".

The discussion then shifted to an area of mutual interest and where Peck thinks there needs to be further research. He questions whether biophilic responses are innate or learned and the ratio of this. I mentioned Kellert's proposal that biophilia is a weak biological tendency and needs to be nurtured. This combines the two thoughts. Peck's other area of interest is the gender difference in biophilic responses and again he suggests that this is an area needing research. Explaining how green Toronto is he said that many inhabitants have great opportunities to connect with green space and proposes that where this occurs in cities their "inhabitants are more receptive to policies that support sustainability than cities that don't have green space".

Peck has had lots of contact with nature growing up. This has been through trips outside the city but also with Toronto's ravine system, the lake and Toronto's forest. He has had a long-standing relationship with nature in the city and considers this probably puts him at the top of the "biophilic health index". Peck wonders what impact the technology age is having on the younger generation. How does the amount of time spent in front of screens affect their biophilic responses? He thinks this is very dangerous, the disconnect that is happening from nature and ecology for the technological generation. "How is nature represented in the virtual worlds they inhabit?" This led to a deeper conversation about how we get more nature in cities, how we keep younger people connected to nature and some of the local initiatives of community groups such as the YMCA.

Toronto had one of the early wide-ranging green roof regulatory requirements in North America. Peck said this took 10 years of work to implement and "the drivers were energy, climate change, and increasing green space". He said the politicians, staff and private sector were all pretty knowledgeable about green roofs and he had been involved with the building of a demonstration green roof at City Hall for which seven companies working on green roofs came together on the project. This led to the first conference on green roofs held in 2003 in Chicago. He thinks that the reason they have done well is that they are "science based and work with researchers plus have involved policy makers and implemented green roof training". Peck mentions Brad Bass who, alongside Roger Hansell, led the research charge and he led the policy and design side. The group was forging a new industry for Canada calling it Greenbacks for Green Roofs. They looked at the multiple benefits of green roofs, having to translate the research from German, as, in 1998, there was only German green roof literature. Peck and his colleagues were motivated by the multiple benefits that green roofs offer. "There isn't a technology that exists that has the range and scope of green roofs. It is the best technology for existing and retrofit buildings in terms of the range and scale of benefits that it brings. There is nothing that comes close to it".

Again Peck suggests that if we are going to redesign our cities or consider the metrics of biophilic cities, then gender response to biophilia needs to be researched.

He thinks gender and design "has a huge potential for further research in biophilia. There are differences in the way genders perceive biophilia".

Peck is also involved in implementing training for net-zero water usage with green roofs. Peck calls the Toronto Deputy Mayor a "champion", as he was intrinsic in implementing green roof policy. Together they worked on standards for green roof construction as well as the regulatory requirement. This requirement also has a financial incentive built in. If a builder does not want to construct a green roof on the new building, they can buy out of it with the money going into a fund that supports retrofit green roofs.

"Different communities have different motivators and drivers. In Austin, Texas, their motivator is usable green space. Tokyo's big driver is urban heat island, where Copenhagen's is climate change and reducing greenhouse gases. Basel in Switzerland's issue is biodiversity because they have mostly wiped out their biodiversity with human urbanisation. So drivers will vary compared with what is going on". There may be alternatives to addressing the particular cities issue but green roof offers multiple benefits. Peck has developed a methodology to calculate the returns that expenditure on a green roof will bring to city (Fig. 3.26).

The interview with Peck was fast-paced, informative and challenging. We debated some aspects of the term biophilic design, which is always healthy and educating. Peck is a strong leader in the green roof industry and has been instrumental in green roof progression in North America.

Vancouver

I was looking forward to Vancouver, having heard a lot of wonderful things about it. We were staying in an Air BNB close to the famous Stanley Park, a sizable nature park with walk and bike trails throughout. Our accommodation had stunning views of the park and the nearby mountains while also being close to the city. My first interview was with a man whose name I had been given by Steve Peck in Toronto.

Fig. 3.26 Stephen Peck, a major driver in the green roof movement (*Source* Author)

Randy Sharp—03/12/2013
Principal
Sharpe and Diamond Landscape Architects

Randy considers landscape architects have been doing biophilic design for 37 years in Vancouver. Urban forest and high tree canopies he puts in this category. In the 1970s, it became mandatory for landscape architecture to be more than aesthetics. Liveability on a neighbourhood level became important. "Where we can't plant big canopy trees, like on roof tops or narrow spaces, hence the green roofs or living walls or other types of greening initiatives. We have the great fortune in Vancouver that green roofs have been part of our zoning for forty years. Basically with high density development virtually all onsite parking has to be underground therefore most of the sites in Vancouver are built landscape over parking structure. When you cross the property line it is a green roof like a conventional roof top garden". Sharp explains that Vancouver consists of a lot of mixed-use development and most of their work is ground-level gardens, intensive green podiums and then the lighter weight extensive green roofs higher up. He explains that "development has to give back to the community, give back to the environment and that zoning responds to form and character". If a new structure is going to be overlooked then the adjacent building doesn't want to be looking down on a grey roof covered with gravel, and hence a green roof would be built. Vancouver doesn't have the same specific green roof regulations as some other cities; in Vancouver, it is largely discretionary. They have the Eco City initiative which sets standards such as LEED Gold, having to earn at least six energy credits and comply with "storm water management, heat island reduction, and green roof coverage hence the drivers for the green roofs". Vancouver wants to be the greenest city by 2020.

In the late 1960s, early 1970s, Vancouver and San Francisco were very close and shared a lot of planning resources. Sharp didn't know the specific reasons but suggests that, like San Francisco, "Vancouver has always had a lot of neighbourhood participation in rezoning plans and local area participation around the City of Vancouver and a lot is driven by community input in overall planning requirements". In the 1960s, there were pioneers like Cornelia Oberlander who were creating rooftop gardens like Robson Square in the CBD. Tomas Oberson also installed an iconic green roof on the Kayser Resources building around the same time. At a similar time, a city planner, Ray Sapksman, was keen on pioneering discretionary zoning and green roofs. So Vancouver has had a long tradition of intensive green roofs, but extensive green roofs have only come into play in the last 10 years. In 1998, Sharp met with Oberlander and an academic Maureen Conelly on a nearby "ecological system island" to brainstorm on how they could put "this pallet of plants or replicate the ecosystems on rooftops in downtown Vancouver". The plants on the island grew in shallow soil so would be suitable for green roofs. At the time, they were importing green roof products from Germany so they started experimenting. Sharp built a green roof on his garage 14 years ago to experiment with the process and roof membranes. In 2003, Conelly had established the Green Roof Resource Centre in Vancouver at the British Columbia Institute of Technology campus as part of the centre for architectural ecology. "She is the cornerstone of the green roof industry here in Vancouver".

Sharp knew of the earlier green roofs in Germany. He had seen the ones in Stuttgart which were built to draw fresh air down into the polluted valley. About 40 years ago green roof companies had become established in Europe. In 1988, green roof technology had just been introduced to North America. Sharp recognised that Vancouver did not have the traditional growing media. What Vancouver did offer though was a wonderful array of volcanic rock so Sharp began experimenting with this as a lightweight growing media on his garage roof with varying plants and media depths. In 2002, he was involved in building the first extensive roof on the coast based on his research. "So that is my story on green roofs".

In 2005, he was involved in a building expansion that needed to remove part of the local rainforest. What Sharp had been doing with his green roof volcanic media was looking at ways to take the green roof and turn it sideways. His approach was to create more of a thermal wall combining lightweight volcanic rock with growing media and plants that grow on vertical surfaces. The creative vision came from a "gorgeous waterfall, but it's really a fern wall with water coming down. It's the inspiration for the living wall. A wonderful vertical habitat, lots of dripping water, a beautiful array of three types of ferns, perennials, mosses. So we wanted to replicate this".

In 2005, Japan hosted the AICHI world fair and the Japanese government had sponsored 30 different green wall systems to be on display so they could export this technology. Sharp attended and looked at some of the systems. He inspired a company from Vancouver to export a modular system and they planted them with 15 coastal plants. From these, they selected eight species to plant in an installed green wall in Vancouver Aquarium in 2006. This was the first modular living wall system brought to North America. This system they considered used less water than the Patric Blanc ones he had seen and could be pre-vegetated off-site plus be able to change modules on site. The wall uses a rainwater system and still exists. "Living walls technologies are largely like living waterfalls. This is one of the concerns expressed around living walls. They use a lot of potable water and a lot of nutrients. The Aquarium uses rainwater but we are looking at using air conditioner runoff and treated grey water".

Sharp has installed another modular wall at Vancouver Airport in 2009. As passengers walked out of the airport, they were struck by a six-storey transit station that obscured the natural views of the surrounding rainforest. The living wall was created to obscure the transit station and merge with the rainforest. "It sits with the overall thematics of the airport, which is to create a biophilic environment that is calming. Passengers typically if they have had a long flight or are tired or are in transit; the airports and the airlines don't want stressed out passengers. So if they take a walk outside, they can see the wall, so they can become relaxed, calm, cool, collected. So the wall gives them an opportunity to not only see this beautiful tapestry back drop but to actually go up and touch the plants. So to be in contact with nature. The airport has been voted one of the top airports in the world for the last ten years. Another important driver for the green initiatives was the soundscape and acoustics". There is also a green roof with a beautiful piece of art from a First Nation carver with the context of the estuary. The idea was to bring a bit of the estuary to the rooftop of the airport. From there you can look out over the area. It is a combination of creating the setting and the story (Fig. 3.27).

Sharp works a lot in high-density urban environments. He has been building rain gardens as part of storm water management systems. He has been building these rain gardens on rooftops and calls them "blue roofs". Sharp also worked with architects to build six different green roofs at Vancouver Island University. These roofs get incorporated into the learning curriculum as a teaching laboratory. "Rain water management is a big driver". The roofs have now become areas for growing and different activities for the students.

Sharp agrees that strong neighbourhood committees and strong planning departments are both present in Vancouver and can work together. "We take it for granted that when we are preparing plans that there is a high portion of site coverage in landscape and tree canopies and roof tops and neighbouring views. People buy properties for the views. Not just long distance ones" (Fig. 3.28).

Fig. 3.27 The green wall on arrival at Vancouver airport and the green roof walkway (*Source* Author)

Fig. 3.28 Randy Sharp (*Source* Author)

The interview with Randy Sharp went beyond our hour as we immersed ourselves in the conversation. He seemed to drift into a timeless zone as he recounted the story of his involvement with the greening of Vancouver. I felt privileged to hear such a historic tale. The emergence story was different to the east coast stories, involving individual champions being innovative in their shared visions and creatively forging ahead with their ideas. I got the sense that community, local heroes and government all share a focus and work together in creating the city they want to live in.

Cornelia Oberlander—03/12/2013–04/12/2013
Landscape Architect O.C.
After I left the interview with Randy Sharp I took his advice and quickly got a train to the airport to see the green wall and landscaping there. The wall was large and beautiful, and the landscaped walks conveyed a feel of the native environment. I noticed the waving native grasses, sustainable, moving and calming. Randy had organised for me to then meet Cornelia Oberlander, a renowned landscape architect whom I had heard of. I went back to the office to meet with her. We didn't have a lot of time but she had brought me some articles and we were able to have a discussion. She explained at the start that because of time constraints this would only be "superficial".

Oberlander explained that she had been involved in "stimulating" the building of a green roof on Toronto's City Hall that was constructed by a company in Chicago.

Our conversation then digressed into who I had been meeting with as she was interested in how I was meeting with people and about my research and thesis. How did I get to biophilic urban design? Had I read *Biophilia*? She was taught by E.O. Wilson at Harvard.

Oberlander showed me photos of the roofs and works she had constructed. One of these was at Vancouver public library. She spoke about the Convention Centre green roof, not one of hers, which is unfortunately having some leakage issues. We then went back to talking about my travels and my plans. When she heard I was going to Portland, Oberlander said I must meet with Tom Lipton. I had been told his name a few times. She said "He got everything going. He is my guru".

Gradually, we got back to her story. Oberlander was married to a very famous city planner and had many children. She was limited in her work but later in life was asked to design Robson Square in Vancouver's CBD. This is three blocks of a green roof.

Her latest green roof is at Van Duysen Botanical Garden visitor centre and has just won a green leadership award at the International Green Roof Congress in Hamburg. An article she showed me wrote that it was a beacon in sustenance and landscape activation. Oberlander wrote an article about it titled "Green Roofs Ought to be Fun". It is in the shape of a native orchid.

Oberlander only works with bigger projects and names and only if she is called in at the very beginning. "I don't take any job unless they are kings. Otherwise out!" When I asked her what had been her motivator, how she had begun constructing green roofs, if it was policy, she replies "No policy, Cornelia's idea!" Asked what barriers she encounters, she says they happen all the time: "Moving along, a barrier. Over that one, move along, next barrier". Then we get back to looking at more pictures of her green roofs. I asked why she thinks people want green roofs: "Well to look down on an asphalt roof is ugly".

Next we get talking about green walls. Oberlander does not like the outdoors ones as "they use too much water". We both agreed that indoor green walls have a value. She mentioned an outdoor one she did like at the Lincoln Centre in New York.

Oberlander says that "years ago she tried to motivate green roof implementation with her book on green roofs". She talked about the slow process in getting one built. I asked her again why she thought they were built and she listed environmental factors: "it cools, microclimatic conditions". Then she read from her book, "Humans have physiological reactions to natural beauty and the diversity of the shapes and colours of nature. Especially to green and to the motion and sound of other animals. This was written by Frederick Olmstead in 1899".

We went back to discussing how in Vancouver it is not direct policy driving green roof implementation but the result of LEEDs certification. "LEEDs is stupid. You don't use water". She works now only on the Living Building Challenge by Cascadia Green Building Council. Here the site is rated as well as the building.

I asked again why she builds these green roofs. "Because I don't want to look down on a gravel roof. And to increase the biomass in the city. And cooling the building in the summer. And warming it in the winter. My library roof after rainfall only puts 28% of the water into the sewer system. So storm water retention and storm water management and reducing the destructive impact of impervious area on aquatic environments".

The discussion went into stories that people had told me of their experiences with green roofs. Oberlander then recounted one of her own when she was rushing through Robson Square, "There is a quiet corner and when I was rushing past a man stopped me. He knew my name but said I didn't know him. He explained that he was a tax collector and found his job stressful. So every lunch time he sits in the quiet corner under beautiful maple trees and goes back to work refreshed".

Oberlander and I discussed the drivers then for biophilic urban design. She sees it as "People. Here in Vancouver it is community and collaboration".

A fellow landscape architect joined us and the discussion led to hospitals and a design they are currently working on. Oberlander had created a garden at a hospital in Vancouver after an accident had led to her being there a while. She didn't like looking out at the asphalt roof, so after her release she created a rooftop garden. Further discussion led to me rearranging my travel plans to stay an extra day and journey with her and her colleague to view the rooftop garden and Van Duysen Botanical Garden Centre.

The offer by Oberlander to show me some of her creations was very special. I knew she was somewhat of an icon in landscape architecture and her colleague informed me she was 92! The wisdom of her years shone through and I could tell she was a leader who inspired with her creativity and innovation. She had said she had done a lot of work in New York as well. Oberlander had brought me one of her books and papers to read. I felt very honoured to get to spend this time with her. She also told of her life and initially being a refugee from Nazi Germany. Her mother was a horticulturalist. Oberlander says she has had a wonderful life, loved her work

Fig. 3.29 Cornelia Oberlander (*Source* Author)

and travelled the world. She has received much acknowledgement for her work, her perception and passion guiding her creativity (Fig. 3.29).

The Next Day
The hospital

I met with Oberlander and her colleague, Ken, at the appointed time and we travelled to the hospital garden she had created. There a senior nurse took us out to the rooftop. Oberlander recounted a story that a patient had told her on one of her visits. "There was a young girl, she was nineteen and she had many skin grafts and it was December. I went to see the garden and she was sitting there but the rain was coming down. She said she loved the rain and it soothed her skin. And then she tells me that when she has finished sitting there she takes a sprig of rosemary with her and then she can sleep well at night". The nurse also recounted how restorative and de-stressing the roof was both for the patients who visited and for the staff. She said a lot of the burns patients were long term but the rooftop had wheelchair access and they were able to come out and plant something in the garden and watch the plant grow. So life goes on. The nurse explained that other health care providers really needed to come out and catch their breath here and often the patients' families as well. She said it was definitely about de-stressing. They had to remove a rose bush as the patients didn't like the fact it had thorns. They were very particular about the plants they liked and wanted ones that were nice to touch; they needed to touch them. The patients love the feather grasses to touch and they like the movement. For the staff, they may suffer vicarious trauma and the garden has become a necessary healing retreat. The nurse explains that "it is a blessed relief for everyone that she can't say enough about. That morning the birds had been singing out there. It's about the sounds too. They have a waterfall and the ability to play music out there. It engages all the senses and it's pleasurable. They are overstimulated in the ward, sometimes can get angry with the pain and the garden allows them the opportunity to come out where it doesn't hurt, it's life affirming. And the seasons! They can see the colours changing and know that the seasons are changing. It's all fluorescent and it's all institutionalised so especially for the stress response, not only for the patients but for the families and the staff and the children".

Oberlander also told of a boy who had watered the plants when he was there and when he left he told his mother he wanted to start a garden at home. We both acknowledged how, given a chance to connect with nature, it can inspire further desire (Fig. 3.30).

Van Duysen Botanical Gardens Visitor Centre

After the hospital, we went straight to the visitor centre. It was wonderful to walk with Oberlander as she explained her design of the landscape and centre to me. As we walked in she pointed out the patterning in the bricks that curved away from the old centre towards the new one so as to lead people there. We admired the natural walls with green cascades and daylighted streams. Then we saw her green roof shaped

Fig. 3.30 Oberlander showing the view to her bed and with staff member in the rooftop garden she created (*Source* Author)

like orchid petals, but also reflecting the distant shape of the mountains. Oberlander said that they had wanted to echo the forms of nature in its design as well as be functional. She emphasised the need to collaborate with others involved such as the architects. Then beautiful designs can happen. When we reached the door of the centre, Oberlander poised and showed me the carved wooden door handle, "very biophilic", she said (Fig. 3.31).

By the door was an inscription–

When one tugs at a single thing in nature, the rest of the world follows

John Muir

Inside it was like we had walked into a mushroom with wooden overhead "gills" comprising the undulating ceiling and leading to a central peak. This also ventilated the space. It was very beautiful and very special. Both welcoming and warm. Here Oberlander brought tea and cake while we admired the view to the outside gardens and talked about the beauty of nature and the wonderful opportunities to create more of this in our urban designs (Fig. 3.32).

We spent the rest of the evening with her until she dropped us at a water taxi. It was a highlight of my travels.

Fig. 3.31 Oberlander at the door of the Van Duysen centre showing the biophilic handle (*Source* Author)

Fig. 3.32 The interior of the Van Duysen centre and view of the exterior green roof (*Source* Author)

Seattle

The train ride from Vancouver to Seattle was very beautiful with snowy vistas. It would be a quick stay but I had two interviews there, one with Judith Heerwagen, a pioneer of biophilic urban design and one of the Rhode Island Group—this was exciting; also an urban designer who worked in planning, providing a good opportunity to hear the policy story of Seattle. Michael Berkshire from Chicago had given me his name.

Dave LaClergue—06/12/2013
Urban Designer
Department of Planning and Development | City of Seattle

LaClergue began by explaining a landscaping factor that they have called the Seattle Green Factor. It is not an incentive program or voluntary, it is a requirement that they have had since 2006. It is a scoring system focused on landscape that counts traditional elements such as trees, ground covers and shrubs but it also counts green roof, green walls, permeable paving, rain gardens and so on. Usually, landscaping requirements in the US have been cut and dried and not really considered other types of green infrastructure. So that is part of what Green Factor responds to. In the early 2000s, people were starting to use some of this "more novel stuff and we were trying to figure out how to give people credit for this if they are doing it voluntarily and how do we encourage people to do more". He then went on to discuss the drivers for this initiative: "First off there was a strong awareness at this point of storm water pollution as one of the major sources of water quality problems in Puget Sound. We are really clustered around this body of water, there is lots of urban density around this water. Over the last decade it has been come really clear that declining aquatic resources in Puget Sound have a lot to do with pollutants that are carried by urban run-off".

Green Factor applies to new developments referred to as urban villages or urban centres. They have a growth management plan in these focal areas of high density. So Green Factor applies to the developers in these areas. When I asked him what group was implementing green infrastructure who the Department of Planning then responded to, he replied, "It was the leading edge developers that were into green building already and trying to do all that stuff". I asked what the initial incentives for the developers were and LaClergue speculated that it was to do with being unique, or elitist and finding a niche. Seattle has had a population boom over the last 20 years, not just with more people moving to urban centres but also due to some big corporations like Microsoft, Amazon and Starbucks moving their headquarters to Seattle. He says that Seattle has gone from a low-density small city and experienced growing pains along the way. "It has created a conflicting feeling among the residents. We have a very environmentally minded population, but being environmentally minded means different things to different people. So some are on board when it comes to urban density as an efficient way to arrange people. But to others it means when you move to higher density you have less trees, you have less landscape or breathing room or open space. So half of Seattle fights the other. I think the Green Factor came about when this debate was under way".

Green Factor is based on similar systems in Berlin and Malmö. "It came about through a company called I-Sustain, essentially an eco-tourism company for policy and design nerds. The company took a collaborative group of politicians, planners, architects, urban designers, developers; this whole kind of ecosystem of different types of people involved in development. This group would take them to Singapore to look at water infrastructure, to look at what people are doing there. Or Berlin, or Copenhagen to look at public space improvements. So there were all these specialists in different fields who went to these cities together and learnt about different cities. Then they would come back to Seattle and have this common vocabulary. Lots had been to Berlin and Malmö so were familiar with Green Factor scoring. So as part of updating development codes in 2006 someone said "Well what about Green Factor like they do in Berlin and Malmo", and everyone agreed and liked how it worked in those cities, so let's try it here in Seattle. There wasn't a lot of debate and it happened in a relatively short time. There was the social capital to do it". LaClergue feels the Green Factor ratio sets the bar much higher for the amount of expected landscaping as well as giving more flexibility in ways of meeting that bar.

Since Green Factor has been in effect, 6 or 7 years, LaClergue has observed a couple of things. Before Green Factor, the typical way a design process for a new development would go is that the engineers, the architects and the developers would design pretty much the whole project and when they were at 90% design, they would bring in the landscape architects to figure out where they were going to put the trees and shrubs. Because a lot of the time it is necessary to do things on the structure to meet the Green Factor ratio, the landscape architects are being brought in much earlier, at the very beginning of the design project, and this seems to be resulting in much better projects. Before, sometimes projects would have "cool" things in the design and then they would get valued and engineered out, but now they have to do it to meet the code.

I asked how the community has responded and LaClergue responds that it is pretty popular, though the initiatives are not always visible. He says Green Factor projects have nicer streetscapes along the sidewalk which are visible, but a nice green rooftop helps "build community" within the buildings by providing a nice amenity to use. Developers either like it or tolerate it because it is a known factor and doesn't add a huge amount of cost to the project. The Planning Department expected there to more resistance from developers. The one challenge has been "an impassioned group of tree advocates who appealed Green Factor as not having to provide trees. We see a lot of trees though as trees are the cheapest way to get credit. So we have never seen a Green Factor project that doesn't have trees. But this group was hung up on the technicality that the code didn't say they had to provide trees. We had to defend it in hearings and so made some small adjustments to recognise that concern. We see this as a very green forward thinking standard".

We recapped what LaClergue considered the major drivers for increasing urban greenery. These were increasing density, increasing population and storm water management. He also mentioned that planning had adopted the Green Factor code at the same time as they updated our storm water code to give more credit to green roofs. This was also at the same time that the "Private market already was starting to provide more green roofs just because there was demand for it. People liked them. It is like a selling point for an apartment building. It is a relatively cheap thing you can do to create an amenities base for the residents. It can't give all the credit to Green Factor. There is evidence that if you are selling or leasing either office space or residential space that looks on to a green roof over a blank roof that you can sell or lease that space for more".

With people meeting their storm water requirements, "they are either going to spend money on a big cistern under the house or visible stuff, and there is a lot of advantage to visible stuff. There are a lot of different factors all steering in the same direction. And it's more attractive".

La Clergue talked about the strongly community driven and maintained gardens that are popping rapidly around Seattle. Seattle is also creating a lot of green streets, where they are narrowing streets and creating rain gardens and bike infrastructure. People want to ride bikes and walk more so the focus is more on creating that aspect of liveability. Unlike Europe, where cities were created before cars, Seattle is trying to retrofit.

I didn't have a lot of expectations in going to this interview but by the end was enthralled. It was a conversation of confirmation of the concepts that had arisen throughout the journey. Here was the lived experience and observation. Again I saw that a major initiative like Green Factor had started with a group, collaboration, a shared understanding. Also because of this meeting point, there was less time and debate involved in implementation. There was resistance only where there was fear and it was misunderstood.

There was clear recognition of the drivers: population growth, density and storm water being the environmental factors, but also good observation of the social drivers, and of biophilic design implemented by the leading, cutting edge, almost elitist developers. It was wonderful to hear that La Clergue had witnessed the building of community through greater provision of green space, such as a green roof. Again the higher property value was witnessed, with people paying more for views of a green roof and to have green amenities. It is pleasing to hear the success of biophilic initiatives.

Portland

I had heard a lot about Portland and was looking forward to arriving. It had a similar feel to my hometown, with an earthy community vibe. Portland had a self-imposed urban growth boundary to protect their hinterland. The river features as a central part of the city and its cleanup has been a driver for change.

Matt Burlin—09/12/2013
Sustainable Stormwater Management
Environmental services
City of Portland

Burlin and I had met at the Biophilic Cities conference so it was good to have that connection established already. I had heard that Portland had a lot of grass roots initiatives and asked him to tell me about these, the top-down policy initiatives and where he saw them integrating. He explained that his area was storm water management, which I thought had been a major driver here, but Burlin pointed out that there were significant others. The river was one of these, keeping it clean, plus neighbourhood streets and developments. Burlin considers that people really understand the relationship between their behaviour and the river and that there are impacts when you make decisions. A lot of people ride bikes in Portland, not all do it because it helps keep the river clean, but there is an awareness of this. As far as Portland being a sustainable city, there are all kinds of initiatives in energy conservation and waste management plus he explains they focus a lot on diversity and equity. Burlin says their top-down policies also really strive to be equitable. They have park systems, natural areas and integrated green infrastructure programs.

Portland is a small city, a factor that Burlin considers can aid implementation of green initiatives. He says the river really attracts people as it runs through the middle of the city so is a central focus in people's lives. "There is an inherent connection. People are thus very aware if the river is not taken care of" (Fig. 3.33).

Burlin also thinks that there are a lot of progressive thinkers in Portland who have been here for decades and this makes the city a bit different. "If we go back to the 70s and 80s there were already people here rattling cages and trying to get

Fig. 3.33 Portland river views (*Source* Author)

land conserved. There were advocates and leaders and I think that kind of leadership fosters more leaders and creates that kind of culture. While we have done a lot in the last decade with green infrastructure and biophilia I think we have a lot of pioneers from thirty to forty years ago to thank for this".

Portland is a very walkable city with a lot of connectivity nodes that not only connect people but wildlife and habitat also. "I think it's a planning philosophy to really look at things comprehensively and strategically; where you have nodes where people can access their immediate needs by walking. Nature needs to be included in those needs. Otherwise it's like saying I have to drive twenty miles to get groceries; I have to drive twenty miles to get nature. Everyone needs access to nature". Portland has an urban growth boundary that increases density and improves efficiency. Burlin feels this also gives the neighbourhoods distinctive character.

Initiatives also sprang from a Federal mandate to clean up the river. Burlin thinks that this was when "the difference was; the leadership, the innovation was when we started suggesting that perhaps we could do this in a more efficient manner if we started using nature to help manage our water sheds and our storm water runoff and prevent sewer overflows". We discussed the recurring theme of the multiple benefits that using nature to respond to a crisis can bring. Burlin then explained that economic perspectives still hinder green infrastructure progress in Portland, yet there has been progress. "When someone makes a decision to do things differently in the world that we live in in the US and it doesn't pencil out as the least expensive option people aren't going to go for it. But I think one of the differences here was that someone along the line decided it was ok to test it out, to see if it worked and then if it does, let's monitor and evaluate it for a few years. And then if it does add up, lets expand it to a bigger area, and then if that works then keep on going. This doesn't happen so much in other places. This takes leadership and innovative ideas. I think what happens then is when you start to integrate these things into the city, and then over time you see the city get greener, then the people who are counting the pennies say 'Oh look, it's helping with our cost of living, it's helping with our property values', and then they are interested".

Burlin thinks this gap from trialling to the potential multiple benefits is hard to quantify. "We don't talk about multiple benefits and equate them to dollars". This is why he thinks that the green infrastructure argument is often lost. In the energy movement, things can be equated to dollars. He thinks that what Portland has seen is an incremental slow evolution towards understanding that these techniques over time will save money. "That's really what we base our whole argument on when we say we want green street policy and water shed protection. We are saying we know this works and we know this will save ratepayers money. When you have an option on the table that offers multiple benefits for the moneys, like green infrastructure, it makes sense to go with that. It also helps make the city more resilient and healthy, yet we still have opposition that say we should use pipes! But things are changing". A budget cut unfortunately cut the green roof incentive that had been in place. He thinks policymakers saw it as a subsidiary and that this is not needed as people should be doing these things anyway. Burlin thinks it was complicated by the financial incentive being implemented in a recession when development slowed down. There were still

Fig. 3.34 Some of the rain garden initiatives Burlin has been involved with (*Source* Author)

135 projects funded though. Now development is picking up but there is no green roof incentive though he is hoping that they will still see green roofs coming in and technical assistance and resources will be offered. “But cost is still a major barrier and without something to help the initial cost for the start-up businesses their costs are going to be higher until there is a broader foundation”.

Burlin explains that although his role is to take care of the sewer and storm water, over time they have realised that this also involves taking care of the water shed; a healthy water shed takes care of the sewer and the storm water. This is part of their green infrastructure system (Fig. 3.34).

Burlin is aware that Portland has a good reputation for green infrastructure but he thinks if it wasn’t for the leaders such as Bob Sallinger, Mike Houek and Tom Liptan, they wouldn’t be doing all this stuff. I had heard of these three and were able to arrange meetings with both Houek and Liptan. Liptan is a retired landscape architect and a big advocate for green roofs who started the whole thing here. Houek is an active ecologist and naturalist plus director of the Urban Green Spaces Institute. “Mike is just a prominent voice and he has done some of the most important projects that have happened here in the last decades, and people listen to him”.

The conversation shifted to discussion of the term biophilic design and Burlin recounted a story around this from a conversation with a prominent Portland landscape architect who was asked about the term: “Why are we calling it that? That is already what we do”. Burlin responded that she was right, that is, what she does as a landscape architect, but it is not what everybody else does. So how do we get other people to do it and understand the language? He is not sure what the right word is but when you are able to say that a street design in a neighbourhood can have an impact on a child’s creative mind, someone needs to explain it that way. “Who do we have to do this?”

I talked about Heerwagen’s thoughts about how biophilic design should not be an add-on. Burlin has been enjoying thinking about how to explain it and expresses this. “I have been thinking about biophilic as a context for our work instead of a new way of talking about this work. It is that we are already doing the work, but now we have a different framework, a broader framework in order to talk about it. For a long time, we have talked about the storm water benefits of a green roof. Then in order to make it sell we would say that it prolongs the life of your roof, you will provide habitat and it will cool your building and over time save money on your storm water fee. But I think what biophilia does is that now it says, 'Oh if you and your neighbours do that then your neighbourhood will go up seven quality of life points.' And it's giving us another step towards saying when your neighbourhood or city is designed like this then your quality of life goes up. We just need to figure out how much it goes up". Calculating the economic benefits may seem daunting at this point but Burlin points out that in many emerging fields being able to do these calculations may always have seemed difficult. Eventually, these things can be figured out. When we get it done, it may get easier.

Burlin discussed how initial decisions made around city design years ago did not take into account, they could not take into account, the impact on future quality of life those decisions had. You don't learn the effects of a city with no green over one generation, you learn it over time. He mentioned the loss of lives in Chicago's heatwave that no one was prepared for, so the decision to take action was vital. Their leadership was there saying "Green roofs, they are going to have to happen. We have to bring these urban temperatures down and that is a quality of life decision that they are making. That makes a lot of sense". Burlin knew of Michael Berkshire in planning in Chicago and the tough implementation time he had gone through, "Michael is a saint!" This is a sign of leadership, maybe aggressive leadership.

I discussed how things seem to be moving from pure function benefits to more awareness of the psychological benefits and Burlin agrees. He sees that in cities which have implemented green infrastructure, results are starting to be seen. Burlin notes that cities like Detroit, a vehicular city, went bankrupt. He sees a new approach coming into cities with people like Jane Jacobs in New York. But questions how much of an impact the cities which are doing well are having, as there are still many who are not doing good initiatives. Burlin is hoping that "as cities evolve it is proved that it is worth that investment in nature for the quality of life, and if you don't make that choice and you go the other direction, then you are going to lose quality of life".

Burlin recounted a story where a landscape architect managed to persuade the property owner to install a green roof. He was hesitant but interested and so they installed it. The property owner loved it and decided to build another apartment building next door and wanted another green roof. The owner loved what this did for his building. "He has space that is open to tenants on the roof. He has created a gathering space and the tenants love it. It really is beautiful".

To Burlin it is about quality of life and healthy living that we want people to be healthier because the city is healthier. He feels that sometimes we can get lost in the detail, or that the idea of a biophilic city can leave out the fact that the individual is healthier because the city is biophilic. Burlin explains that his job is to keep the river clean, but this is not the bottom reason, it is so the river is clean for the people, the fish and the wildlife so they can be healthier. He thinks there should be an office for the quality of life, which sets strategies to increase quality of life points. We need to be able to value quality of life. Does biophilia do this? "There is no real context for quality of life, there are so many different ways to make your life better but are we really valuing this as quality of life? Like trees are good, clean air is good, air quality is good but it is not an integrated approach to improving quality of life. All these things fall under the same umbrella yet in some way need to coalesce. That is what I like about Biophilic Cities. This is a way to build them so that all things make sense".

Burlin says that The Portland Plan is starting to incorporate new things, like trees and green streets. He reflects that our cities have been designed with such different focuses yet is confident that if we built a city from scratch today it would be built green and sustainably, it would make sense. "We are having to retrofit now". Portland generally shows much support for green initiatives and campaigns to protect natural areas receive a lot of community support. It is a young community that embraces being outside. Many of the inhabitants have moved to Portland so they have greater access to nature. Sustainability seems to be part of the culture. Green building is big in both residential and commercial. "What has drawn young people to Portland is the nature and so it makes sense they want to care for nature and incorporate biophilic initiatives. It is a vibrant energetic city that takes care of its natural resources and people appreciate that".

We discussed how once a certain mindset develops in a city then it tends to attract more people wishing to live that lifestyle, so it ripples out and grows. Then there are some strong community advocates who connect with the government. Burlin explains that usually where there are push backs to the green programs it is in areas where there are greater problems and need. These are the more disadvantaged, outer suburbs with higher crime, infrastructure disrepair and dirt streets. Ironically, it is these areas that need biophilia the most. So sometimes sustainability is a luxury and implementation can be a little skewed. "When we come up with the road map for biophilic urbanism, we have to come up with the equity for all communities. We have done a lot of outreach in the less advantaged communities. We need to approach it so we are making the whole city healthy, not just parts of the city. Quality of life for everyone is quality of life for everyone" (Fig. 3.35).

The conversation with Burlin was extremely thought-provoking and informative. The recurring themes of leadership and innovation were reinforced—a group of

Fig. 3.35 Matt Burlin and Burlin on the City Hall green roof (*Source* Author)

leaders with a common vision supporting one another. I got a sense of the interface between community progressive thinkers and government policymakers. Innovative, strong local champions had a voice and influence. Burlin also conveyed the sense of the community that is drawn to Portland because of the access to nature so there is a mindset that develops and ripples out attracting similar thinkers. I think this helps make Portland special.

Burlin conveyed a depth of perception and deep thinking regarding the issues of biophilic urbanism and it was stimulating to discuss these. The multiple benefits that come with implementation have arisen in most interviews. Burlin explored this further, discussing the difficulty of making the economic case and being able to quantify the benefits. Maybe this is where biophilic urbanism sits; a framework for integrating the benefits and outcomes of a multidisciplinary approach to urban design. An integration that I think really needs to happen.

Burlin's thoughts about it all coming down to quality of life for all were insightful and they really did illuminate the bottom line of what is trying to be achieved in biophilic design. We need a greater understanding of the broader picture, the deeper perspectives and recognition of the need to coalesce into a more integrated framework for quality of life.

Burlin also offered to show me around Portland and the next day he took me to some of their natural parks, the green roof of their City Hall and showed examples of their green streets initiatives and kerb-side rain gardens. It was generous of him and wonderful to see these examples.

Tom Liptan—10/12/2013
Ecoroof Program Manager at City of Portland
Portland, Oregon Area
Environmental Services

I had been hearing about Portland and was told I must visit during my journey since they were doing a lot of green initiatives there. Arriving in Portland I immediately sensed a similarity with my home town of Fremantle. Our accommodation was with a family in a beautiful neighbourhood close to the city. It had painted roundabouts and they had a chicken coup with a green roof in the backyard! I was to have my first meeting in a nearby café which I could walk to.

I had heard about Tom Liptan from a few people. He was renowned as a pioneer, innovator and leader in the area of green infrastructure and biophilic urbanism. Although retired now he was willing to meet with me and show me around Portland.

Liptan took me for a drive to show me a particular apartment building development which he was involved in and which illustrated good biophilic initiatives. There were storm water issues at the site and an underground stream water pipe that needed replacement. This would be expensive for the developer and Liptan had discussed options with him. The outcome was the decision to 'daylight' the stream, bringing it to the surface and making a feature of it. There was also a triangular intersection of roads outside the apartment block and the stream continued under these. Previously, a development had created the third side of the triangle as a perceived 'improvement' but Liptan pointed out that it had actually had been a detriment to the area plus it was

Fig. 3.36 Tom Liptan with the reclaimed road and the daylighted stream (*Source* Author)

unnecessary. Part of the apartment development project then had been to remove the road and create rain gardens and a stream bed to manage the water instead (Fig. 3.36).

I asked him why the developer had chosen to daylight the stream instead of replacing the pipe. Was it due to economics or aesthetics? Liptan replied that it was both. He then suggested that "it was the three 'Es'. Economics, Esthetics and Environment". The developer liked the idea that he was going to bring to life a stream that had been buried for over 75 years. This added to the economic benefits. The developer also spent extra money to daylight the creek through its full course and make it a feature. This was at the cost of extra apartments but made a better project. The outcome was "higher rents and longer term tenants".

I had heard that Liptan was responsible for the first green roof in Portland so asked him about that. He confirmed that intensive green roofs had been installed but this was the first retrofit of an extensive green roof. This was his garage. "I was probably the first person in North America to monitor the rain and the run-off, or lack of run-off, from what we called eco-roofs. This was in 1996. It worked! It was growing and then we had a big rainstorm and I was out there with measuring cups and found it worked very well". At the time, Liptan was working in storm water management in government and had heard about green roofs and thought: "It just makes sense. If you have soil and vegetation on the roof it is going to absorb a certain amount of rain and help with storm water management. If it does then wouldn't that be great for the City of Portland to use as a method to help storm water run-off getting into our combined sewer system". I asked him where he had heard about green roofs. This had been from a London-based man from the London Ecology Unit who spoke at a Portland conference on the benefits of including small natural areas in urban London to educate children. In the informal conversation outside the presentation, he had told Liptan about the concept of green roofs and that this was something his organisation was looking into. The presenter said that some big projects had happened in Germany and Switzerland. Liptan was intrigued by the idea and thought "Wow, that sounds like it makes a lot of sense. He also mentioned something about water management so I thought this was something I needed to find out more about". A couple of years later a book came out from this London group and he bought a copy. He was still thinking about this when his wife bought some detergent that, on its label, said it came from a factory in Belgium with a grass roof. Liptan called the number on the label and they sent him an eight-track video of the opening of their green roof. He then decided there really was something to this and he needed to find more information. His boss agreed and Liptan was assigned to this research. He started collecting little bits and pieces of information from Germany. This was happening in the mid-1990s.

The question arose though: "So they do it in Germany, so what. How do we know it works in Portland? That was what everybody asks. So I need to trial this. I need to find out. I have my old garage out there". In January 1996, he constructed the green roof on his garage. After the success of his trial, his employer, the government Bureau of Environmental Services, decided to do another test as a full-fledged project. The City Commissioner was on board and there was some publicity around the project. They discovered a developer constructing a new building who was interested in providing the rooftop and it was funded by the Bureau. It was called the Hamilton Apartment Building and completed in 1999. Monitoring equipment took 3 years to install and information began to be collected in 2002. They monitored rainfall, runoff and speed of runoff. They discovered runoff was 90% slower than from a conventional roof. This translated to volume control and peak flow control. Water quality was also looked at and Liptan explained the complexity of examining water quality. Water is filtered by the roof but also may pick up nutrients. So where the water ends up is important; into the ground is good, into a water body is not so good. But generally the water was fine. Temperature was also considered as warmer water can impact on their cold water fish.

The trial was considered to be successful and provided a viable option for storm water management in Portland. "It was regulatory to manage storm water and a green roof provided a choice on how to do it".

Liptan thinks that a roof of a certain size should be a green 'eco-roof' and at the moment knows there are people working towards putting this into policy. "It just seems like how long do we have to wait? Toronto has passed an ordinance. We would be the first city in the United States. No one in the US requires it, it is all based on incentives". We discussed Chicago and Liptan explained that it is not in their code, it is under conditional use. From my Chicago conversation, I recalled that it was in their Sustainable Development Policy which developers had to adhere to if receiving financial or zoning assistance. So Liptan was correct. Currently, Portland has incentives. Green roofs were first used to provide a choice in managing storm water for developers. The concept was then put in as a density bonus, which gave two incentives for developers to use green roofs. A financial bonus was brought in, five dollars a square foot for a green roof, but this only lasted about 4 years until a change of Mayor.

Liptan had also thought about the question of why Portland has a 'green' reputation with some greening strategies, such as green roofs, being initiated here. He considers it is people like him "who think something is good and pursue it and get it going.

His motivation was considering whether it would do the job he was working on". Liptan is a landscape architect not an engineer but he was working on storm water issues usually addressed by engineers. "When you start adding different perspectives to a question, you start getting different approaches". His training varies from an engineer's and he approaches utilisation of space in a different manner. Liptan was the only landscape architect in government working on storm water issues. "So for me it was why don't we manage it with the landscape? Why do we have to put it in pipes and stuff? Engineers don't touch the landscape, it's not their job so I said "Why don't we touch it together?"

The conversation turned to the idea of the drivers of biophilic design coming from risk-taking champions. Liptan commented:

> It seems like that is a phenomenon of human species and perhaps there are more people in that category in Portland than other places as it relates to the idea of the environment. I don't consider myself a champion. I just consider myself as a guy who saw something and then brought it forward. But maybe a champion is someone who if they get ridiculed and you get pushed around and you keep talking then you are definitely striving towards something that others are trying to put down. People used to laugh at me. They would think I was just a whacko guy when I talked about eco-roofs. This one guy said to me recently; 'You know, Tom, I used to think you were a nut but now I see you were probably right.'

Liptan thinks it is the responsibility of people in government not just to do what the job says but to work for the citizens of the community. "It's hard because your job is about doing the ABC, you need to do your job. So I was questioned as to why I was doing this green roof thing, but if you see something that can make your community better then I think you need to bring it out. If people at the high level have this attitude then it makes it easier for people like me at the lower level".

I told Liptan the story of the real estate woman in Washington DC who had determinedly produced a green roof on her building. He supported this, suggesting that it doesn't have to be someone working with the city, that anyone can do this. Liptan says there are Portland stories of private individuals seeing examples of biophilic design, thinking that it makes sense and they just 'get it done'.

We discussed the multiple benefits of biophilic design and how it crosses into many disciplines and attracts different people. I mentioned that many of the people I had spoken with at some point would say "it just makes sense". Liptan responded instantly with some thoughts:

> You know, what strikes me about part of what causes the regression of good ideas and good practices that cities have is actually the lack of understanding of what the ramifications of something new will be. People can sell a product that does the job, but if you are not informed enough about when there is an alternative to the product, you will often buy the product. Like the extra road outside that someone decided needed to go there when it didn't. And people use their money to promote their product. Economics and consumerism, we don't want to get rid of it and we don't want to overly control it because we know that doesn't work. So people need to get good advice or do their homework. Buyer be aware! And learn when you don't need to be buying something.

Liptan sees this thinking as impacting on the way we design our cities. He spoke about when he first heard Tim Beatley lecture in Portland on Biophilic Cities. Liptan wondered which city in the US or in the world is the model for a biophilic city. Beatley showed bits and pieces from cities but not a whole city. I mentioned Singapore, which led to a discussion on Singapore's system of governance which imposes a lot of these initiatives. Liptan wondered if this "strict green control of the environment" would be the only way that humans will actually do biophilic initiatives. Liptan was very reflective in grappling with this topic as he considers people do not like being told what to do or not to do. He thinks we need something other than regulations; understandings of what standards are important to have and which are not.

We discussed the 'ripple effect', where people see the proof, the "show me the money" as Liptan calls it, and how this can work in implementing initiatives. Liptan pondered further:

> I wonder if we, us human beings, need to make so many mistakes before we get to so many successes. Is it just a natural process for us that we make mistakes and then try and correct them, generation after generation? My limited sense from meeting young people is that many of the younger generation are asking questions differently to the way the older generation asked questions. I think this is somehow related to religion. Whatever it is, how does this play into so many things we do as a human being? Spirituality is a very fundamental part of our thinking, and our beliefs affect our thinking. People look to their beliefs for the answers to things they can't answer. It seems to me young people are looking differently and this is encouraging to me. They are not automatically accepting of some of the things we do. There is something positive about the younger generation that is encouraging to me about the future of the earth.

Liptan was a reflective and deep thinker and that led to a fascinating conversation. Again I felt I had encountered a special wisdom with an altruistic approach to life. I wondered if this contributes to his creative openness and innovation. The converging events that led to his garage green roof trial had translated into such a serendipitous journey, even involving a detergent packet! How long it would have taken Portland to get a green roof if Liptan had not seen that packet?

Liptan is a risk-taking champion, a local hero, in my view. He stood up through ridicule for something that made sense to him and that did the job he was assigned to do. The fact that he was a landscape architect in a traditional engineering area introduced a different perspective. Perhaps if we did this more often with problems then we may find more innovative solutions.

As he showed me around Portland we talked in more depth, with philosophical conversations about life and about our cities. Tom considers that initially we had to modify the environment to provide our needs, like shelter. We just forgot when to stop and the economics of sterile, right-angled designs took over. We both agree it is time, and necessary, for humans to reclaim our connection with nature.

Mike Houck—11/12/2013
Director
Urban Greenspaces Institute

By the time I met with Houck I had heard much about him in previous interviews. He was spoken about as a strong community activist who had played a major role in the development of Portland.

Houck has always been interested in birds and nature conservation. He had been involved in protests about filling in wetlands near the Portland River. Some years later, as he was driving into the city from a bird-watching expedition, he had "an epiphany":

> I found myself getting uptight about coming back to the city. I think primarily because cities have been demonised and I kind of bought into that stupid view of cities. So I actually had a conversation with myself and challenged myself about the good things I like about cities like going to the theatre etc., etc. Shortly after that a friend handed me proceedings called 'Wildlife in the Urbanising Environment'. The authors were doing research on the importance of green spaces in urban environments. I read that and it triggered in me a desire to start focussing more effort on nature conversation in the city. This made perfect sense.

Houck was approached by Oregon Department of Fish and Wildlife who he said were not very interested in urban wildlife as they didn't think there was much. They offered Houck a small grant to take on wildlife in the metropolitan region and, as Houck didn't have a job at the time, he said of course. He took the title Urban Naturalist. One of the goals of the urban growth boundary legislation is for an urban wildlife inventory and this is what Houck did. He did this pro-bono. This led him to being very engaged with protection of fish and wildlife habitat in the city. It had a rocky start as the local planners told him there was no place for wildlife in the city and the idea of the urban growth boundary was to protect wildlife outside of the city. "This has been a protracted battle for the last 33 years. But in the last 5–7 years there has been a corner turned and local planners have acknowledged that it would be nice to protect some nature in the city both for the ecological value and to provide access to nature for the people who live in the city".

Houck felt this wasn't getting much traction so developed some tools, applied for grants and researched overseas. He received a significant grant to develop a refuge system for urban wildlife. This Houck considers significant as it was a shift in strategy from planning to a non-regulatory approach. Metro is Oregon's regional government with jurisdiction over the cities and counties. By law every city and county has to amend their local plans to conform with the regional plan. This is unique in the country. Houck hoped they (himself and Tom Liptan) could implement change through convincing Metro to do it differently and become the owner of natural areas. Houck first convinced county commissioners to divest themselves of all their property and give it to Metro. Houck then convinced the Metro council to adopt a region-wide master plan for parks, trails and natural areas. In 1995, citizens even voted to tax themselves to provide money to Metro to buy more natural areas. Metro has acquired 16,000 acres of natural areas to put into public ownership for protection and appropriate public access. Simultaneous to this, several symposia were held called 'Country in the City' which brought experts to Portland. Storm water management experts attended one of these, so Houck explains that this was another agency who joined in the growing collaborative partnership. "Now we have Fish and Wildlife agencies, Park agencies, the Regional Planning agency and now we have storm water. So we are building the partners all of whom make a significant contribution to the whole, to the system".

Houck then talked about the NGO Coalition for a Liveable Future group he had helped to initiate. One of the objectives was to create regional policies and regulations to better protect urban wildlife and habitat. These were presented to Metro and adopted. He summed up saying how they had gone from regulatory to non-regulatory acquisition and back to regulatory. I commented on how he had managed to get around any blocks and he agreed, saying that "we had to be very creative and willing to switch strategies as we went along". Houck and his peers have published books on wildlife in the city and tried to build political will and public awareness. "We run field trips and still hold conferences and symposiums to share research with others".

Another group Houck has been a prime driver in forming is The Intertwine, a coalition of public, civic, private and non-profit organisations who are dedicated to building a world-class system of parks, trails and natural areas. The Portland Regional Conservation Strategy is a product of The Intertwine Alliance (Fig. 3.37).

We went back to the earlier concept that had intrigued me, the fact that taxpayers had voted to tax themselves to pay for land acquisition. Houck then stated that even more astounding was the fact that last year (2012) they had voted to tax themselves to pay Metro to manage this acquired land. Houck made another very interesting point that some of the "resistance to the biophilic initiatives came from the same people who are acclaimed for making Portland such a green city. Because what they focus on is light rail, land use planning and to hell with nature. Portland has a great reputation, but quite honestly the green stuff started just recently, it's been a fight. The power structure viewed nature as outside of the city".

Houck does see that over time there has been a shift in the community attitude as he would never have had the support he received in the later years. He thinks that people have started to care about what happened in their backyard and have become engaged. "There are now too many people to count. Yet I think even when I first started out people did care deeply about the wildlife in their back yards. We were just a vehicle for them to express their interest".

I mentioned the 'green' reputation that Portland has. We both thought that this would tend to attract people seeking this and so make this aspect stronger. Houck pointed out that green does not always mean biophilic and caring for nature: "When I first started this, government agencies were not supportive but over the years we have developed a good working relationship where I can beat up on them on Wednesday but on Friday we are having a beer. There is now a level of trust". He is concerned about this 'cohort' retiring and the new relationships he may have to develop.

The connection between human health and nature is recent but it is very explicitly part of what we are doing now: "There is more of a focus on this human-nature connection. I still take people out into nature and last weekend a flock of snow geese flew over. People are interested in the critters but I think they were more moved by just the beauty of them flying overhead".

Houck only became interested in nature as a graduate student. A backpacking trip into nature turned his life 180°. "Maybe I am like a religious convert, I am more zealous because I am a recent convert. I will acknowledge that I have had a huge impact in Portland, but what has been really most rewarding is the relationships that

Panorama

inside

28 BIRD'S EYE VIEW
PDX ECO-IDENTITIES

30 ON THE RISE
A NOSE JOB FOR TIMBERLINE LODGE

34 ON THE TRAILS CAMPAIGN WITH DAVID BRAGDON

36 TAKE 3
LIVING LARGE IN LITTLE SPACES

40 HINDSIGHT
A BILLBOARD FOR THE BIRDS IN OAKS BOTTOM

Houck's First Stand

Many a visitor to Oaks Bottom Wildlife Refuge might wonder why a 70-foot-tall blue heron is painted on the wall of the adjacent Portland Memorial Mausoleum, standing guard over Portland's most visited swamp. But an equally intriguing question would be: Why is there a 140-acre wildlife refuge here at all?

The key to both mysteries lies in the hands of urban naturalist Mike Houck. More than any other individual, Houck has led the charming, cajoling, and arm-twisting necessary to add 11,000 acres of protected natural habitat to the metro area in the past two decades. His latest undertaking is a 50,282-square-foot wildlife mural that will cover the mausoleum's entire western face—though some fellow activists, like Metro Councilor Robert Liberty, believe the wall would be more fittingly filled with "a 60-foot-tall portrait of Houck."

see full story, page 40

photograph by Daniel Root

Fig. 3.37 Houck—a local champion (*Source* Author)

have evolved from the time there were three of us working together to now, when there are more people than I can count".

Houck is a true local champion. His concisely told story, which started with his personal epiphany and then related the initiatives and milestones he had achieved, was inspiring and illuminating. He accepted the core biophilic principle that nature needs to be part of the city and not just outside it. Houck was an early adopter of this, arriving there through his own inner reflections. Sometimes it seemed like Houck had had the right things come along at the right time, but mostly it appeared his passion and strength of convictions had led him to find ways to make it happen. He also mentioned the relationships he formed along the way that had supported him.

Houck's story is a wonderful illustration of where bottom-up can meet top-down. It clearly demonstrated how informed individuals can make it happen and can influence policy, similar to the thoughts of Mary Rowe in New York. But a big part was the cleverness of seeing the agencies that had an interest, of bringing them together and creating a synergistic collaborative outcome.

Ada Karlos—10/12/2013
Homeowner

Ada and her family were our wonderful Airbnb hosts. On our arrival at their home, I was happily surprised to spot a green roof on their substantial chicken coop. Ada agreed to talk with me about this.

The roof is 7 years old and the idea came after Ada had researched into what makes a comfortable environment for an urban chicken. Portland can have extremes of hot and cold and Ada sees "this as a way to control the temperature. It never gets really hot inside". The roof is also visible for their kitchen window and this is something Ada gets pleasure from. When she arrived in Portland Ada became aware that a few other people had small garden roofs but she thought it would be good to encourage more. She visited the nursery to find out which species would be best. There was a lot of interest in the roof and it was picked for the annual community chicken calendar and became part of the Tour de Coop, where people visit interesting coops. So far Ada has had a couple of hundred visitors to the green roof (Fig. 3.38).

Reflections on Portland
Portland is unique. It has a reputation for being green, for being community oriented and active and for being very liveable. I felt like I had interviewed three local champions. These interviews really illustrated the meeting between community activism and government agencies where bottom-up meets top-down through an interface of information, collaboration and persistence. There was creativity and innovation plus serendipitous events that all collided and were supported by three main players who together changed policy and the direction of Portland. The encouragement and protection of urban habitat may have come later through someone else, or may never have happened. These early adopters and innovative thinkers, whether in civil society or government, seemed to play a significant part in the emergence and coalescence of biophilic initiatives early in the biophilic design social movement journey.

Fig. 3.38 The green roof chicken coop (*Source* Author)

San Francisco

San Francisco offered a few interviews. Two major players had been at the Biophilic Cities conference and one had emailed colleagues and organised some other interviews for me. Both these conference attendees had presented so I had some background on initiatives in San Francisco. This city was climatically quite different from others I had visited and therefore had a different agenda. Storm water was not the issue it was in most of the other US cities. Judith Heerwagen had also provided me with a contact at Google. Both Judith and Bill Browning had been involved with their architects in designing biophilia into the Google headquarters at Silicon Valley. I was excited about this interview as I was told it is not so easy to get to visit Google headquarters.

Peter Brastow—16/12/2013
Senior Biodiversity Coordinator
SF Environment

We began the conversation by discussing the term biophilic design. Brastow considers that there are similar terms but maybe they are not as encompassing as biophilic design. Brastow's career has focused on restoring habitat in the city and connecting residents to this habitat and nature in the city. He considers that this "is so they can deepen their sense of place and is also for the sake of nature itself. We have some amazing habitats here and biodiversity inherent to this particular place. It's about recognising that just because we are a city doesn't mean that we shouldn't take care of the biodiversity that is indigenous to and special about this place".

Brastow considers the human–nature connection as "fundamental. It's like a mutually beneficial relationship and process. Nature needs us to heal and manage and steward." He works with voluntary community stewardship where people have a stake and restore the relationship. We need to encourage people to be engaged and reconnecting "because it's really about restoring a whole new relationship with the earth". He recounts the number of stewardship programs around the city in the city parks where people are involved but also expresses that "we want to expand this to where we are really bringing nature into the built environment so that the built environment becomes more compatible with our remaining natural environment and so ultimately we are evolving towards a more wholistic blend of the two in place". There are some public funds but Brastow thinks more tax money should be allocated to take care of the nature of the places where we live. This is a matter of priorities.

The stewardship programs in San Francisco have blossomed in the last 25 years. There is a huge need for more people. One barrier encountered by Brastow are the unions "who are not always super happy about using tons of volunteers. But it improves the volunteer's quality of life".

I commented on the number of little parklets and small gardens I had seen around San Francisco. "Yes. There is a real movement of people taking over little pieces of park where they have been permitted to do so, or derelict areas, and greening them. It's part of the street parks program where you can apply to do this. People adopt little places. Some of these have been connected to provide restorative biodiversity corridors".

Brastow had founded a non-profit organisation called Nature in the City in 2005 that is also involved in the parklet management. It is still successfully functioning. He had worked with the National Parks service early in his career and it was this work that got him interested and led to the idea of creating an NGO. Brastow wanted to be more in the heart of the city and focus on urban nature and biodiversity. In the early 1990s, there had been recognition that there were still some significant natural areas in the city and that a program was needed to take care of them for nature and biodiversity. This occurred with the hiring of a biologist and community supportive activism. Some sectors of the public provided quite a challenge by "being cynical about restoring nature in the city. Nature belongs outside of the city in parks and people are in the city and the parks are for dogs, they would argue. And others wouldn't want us to cut down any trees, even when it was necessary. These cynics and skeptics and critics made nature in the city sound kind of negative".

Brastow considers there are four goals that are a focus of San Francisco environmental agencies.

- Restoration,
- Nature in the City,
- Stewardship and
- Environmental literacy.

He thinks that agencies need to be more "knitted together" to achieve these goals. To help achieve this he has been involved with creating an inter-agency biodiversity working group who are trying to collaboratively achieve these goals. "All these different agencies have a stake in this whether they own land, or whether they own the street-scape or the utilities commission and the storm water stuff. So the plan is to get the whole city family on the same page and speaking the same language so coming up with a unified narrative and unified approach to land management". This would give the public better and more unified information on which to base decisions.

Brastow understands the need to bring all the factors together. Not just "do the green"" but consider the water use and the biodiversity. He would like to see green roofs created which do not need to be watered so much through summer. The ones Brastow has seen require a lot of water. "We need to experiment more. Roofs are green roofs in winter but in summer they are brown roofs, but they are still alive! I want to also see sustainability and water conservation". On the east coast, it's a lot easier to green roof as it's raining all the time. The roofs are green and then go brown in the cold of winter. In San Francisco it is the opposite, roofs go brown in summer.

To Brastow, connecting to nature where a person lives is connecting to taking better care of nature around the planet. "Cities are fundamental to environmental sustainability. If we connect people to nature at home then it ripples out and they relate to all the other environmental stuff all around the world. We need to get people outside with a unified voice. Stories can get people to key into nature in the city".

I found the conversation with Brastow refreshing. He worked in city government but the focus was not about managing an urban crisis through nature, it was about nature itself, biodiversity, and connecting people to nature to increase their sense of place and land stewardship. There were similar themes to Portland, such as the community group who thinks nature should be outside the city. Similarly also was the recognition of the need to bring all the stakeholders involved with nature in the city together in collaboration that nature in the city bridges many aspects of city life and city design. I think the idea of voluntary stewardship that Brastow works so hard towards is potentially a powerful tool in changing the relationship of urbanites to their nature in their city. Brastow conveyed well his understanding of the movement in our cities towards creating a new relationship with nature.

Scott Edmondson—16/12/2013
Planner/Economist
Information and Analysis Group
San Francisco Planning Department

I met with Edmondson directly after my meeting with Brastow as they work in the same building. We began talking about the Biophilic Cities conference and the focus on biodiversity. "It was where people's energies were at". Edmondson then pointed out that some of the conference was focused on economics but "it begs the question of how this stuff is related". The most advanced sustainable planning systems for the built environment, says Edmondson, are the Living Building Challenge and the Eco Districts. "They both have biophilia as key components of their frameworks, but the way that they define it is kind of rudimentary. It's just nature in the city. It's more traditional".

Green roofs in San Francisco currently are created through the storm water management program. Biodiversity is seen as an element of sustainability but "sustainability is such a fractured kind of unsystematic, un-strategic initiative these days. Everybody is doing bits and pieces where they can so it's not stitched together".

The conversation then turned to aspects of sustainability. How dependent should the human economy be on the natural economy? "There is a relationship. Right now the human economy is systematically destroying the natural economy. That's the sustainability problem. No matter how much biodiversity we create if our economy continues to destroy the natural system, then we have a problem".

Edmondson, being an economist, naturally has an economic slant on biophilic design and the relationship between the human and natural economy. He discussed the necessity to link these together in biophilic design, to recognise the impact and interrelationships of one to the other. San Francisco has both an Urban Forest plan and a Green Connections plan, both of which "are plans to integrate nature and habitat value into the human environment".

Brastow had talked a lot about butterfly species found in the San Francisco area and one of his stories was retold by Edmondson with a different perspective that I found interesting:

The Western Tiger Swallow Tail Butterfly story

> In one of the major streets of the CBD there was a recent discovery of the Western Tiger Swallow Tail butterfly. 'Tigers on Market Street'. So that is a big phenomenon. But what is really interesting about it is that it is not so much biophilia by design, but it is nature invading and adapting to a human space. Adaptation is occurring; those trees are not the right trees in terms of species, but they are laid out in a linear way which is how the butterflies use them. Also in Market Street the trees are in the bottom of this big canyon, which is like a river canyon. The hypothesis is that they are adapting because there are those similarities. It was also discovered that these butterflies and some other species were congregating in the road intersections. So what we are seeing is that what you would think is a total 'a-biophilic', 'a-nature' environment is being reinterpreted and being re-understood to have characteristics that mimic a natural environment. So this is the reverse. It's not us doing biophilic design, it's nature adapting and illuminating the biophilia. The redesign plan on Market Street had been including ripping the trees out and it was only the discovery of these butterflies and intervention by an activist group that stopped this. Now the plan is to look at how to redesign Market Street to include the butterflies and other animals. This is biophilia in action.

We discussed how people respond to butterflies. Edmondson thinks there is a special connection people have with butterflies and people are very keen to protect them. We both wondered if it was because they are so beautiful and it triggers people's response to aesthetics.

Although there is not a focus on biophilic design on Edmonson's planning department, he is aware that there is a growing interest in the city to bring San Francisco up to speed on green walls and green roofs. A policy report has been written that summarises other cities' experiences and contains relevant information that would go into formulating a green roof, green wall policy for San Francisco. This had followed from the Planning Department sponsoring a *CitiesAlive* conference in San Francisco in 2013, which had generated significant interest. At this point, there is nothing in the building code or planning code to implement green roofs, but a green roof can be used as an option to manage storm water. Edmondson doesn't think there have been many takers for this option. It is all still under debate and unclear which department would be undertaking any green roof initiatives as far as policy. Edmondson thinks there is more of a drive to do biophilic design among architects and landscape architects.

Fig. 3.39 Scott Edmundson showing biophilic features in suburban San Francisco and Scott with son, Jett Jackson (*Source* Author)

Architects have a big capacity to influence the design of a building. The *CitiesAlive* conference had brought that thinking out into different realms of planning and water management.

The conversation concluded with a discussion about economics, which Edmonson sees is an important dimension. "The biosphere is the principal economy of the planet and the fundamental principles of this regenerative system or regenerative economy are the ones we need to mimic in our human economy. The transition to this is the social system, economic system design challenge" (Fig. 3.39).

The conversation with Edmonson was again a different perspective. I was speaking to an economist. This did make me think and consider things in these terms. I liked the idea of viewing things as interrelating economies, the human and the nature. It also though is a matter of terminology as the principles are also ones of sustainability. I wondered if the Tiger Swallow Tail story represented the perspective of self-organisation from the natural world. It reflects the same principles that Mary Rowe in New York talked about: self-organising and adapting for survival.

Kirsten Weeks—19/12/2013
Energy and Building Ecology Specialist
ARUP

Weeks has always wanted to make a big difference and "do as much good as I can for the environment, with people as part of the environment. I explored different areas and ended up in design, particularly green buildings and sustainable design. I find there is something particularly compelling about green roofs and restoration of natural ecosystems in the built environment; and then also using either actual natural ecosystems or using built systems that work like natural ecosystems in order to solve problems in the built environment. I design built systems that take inspiration from natural ecosystems in that they don't generate waste, they use waste as resources, the loops are closed".

Her passion for green roofs has seen Weeks involved in efforts to set the stage for a lot of green roofs to be built. She has observed that there has been an increased interest in San Francisco in urban agriculture, biodiversity and green roof policy, particularly in relation to storm water management. Weeks instigated meetings with interested parties one of whom was SPUR, a San Francisco policy and urban design think tank. The SPUR green roof task force, of which Weeks is part, compiled recommendations for green roofs in San Francisco which were presented at the *CitiesAlive* conference and endorsed by a San Franciscan 'supervisor'. Supervisors are government representative who sits below the Mayor.

Soon after Weeks joined ARUP she was involved in a study with ARUP and Walmart who had approached ARUP to study green roofs. She thinks Walmart was interested because they thought that a green roof could save money and help them reach their energy and environmental performance goals for their buildings. Weeks has also been involved in a study assessing the green roof and policy environments in different US cities. The question the study addresses is how much it cost to build a building and meet policy, with and without a green roof. She explained that in this perspective the "premium for the green roof shrinks a lot and then pays back in a shorter time". This study then led to doing a study for the United States General Services Administration which commissioned ARUP on behalf of Congress. This was to research the benefits, challenges and cost performance of green roofs on public buildings. In doing the study, they put "dollar values to the benefits, including the social benefits and things such as air quality and habitat value and heat island reduction. When you added it all up it did have a positive value for the individual owner/occupier but not a very big one, about $2.60 per square foot. But to the community it was a significant $30 per square foot. So we could take this to the policy maker and say you are quite justified in offering an incentive of say $5 a square foot because the community is going to benefit six times that".

> I do this because I believe inherently that this is a valuable thing, to cover our hard surfaces in our cities with something that mimics a natural ecosystem. But as I have gone through these benefits, I feel my hunch is substantiated by research.

Weeks understands that sometimes a green roof on its own is not the most cost-effective for a particular outcome, but it is the multiple benefits that make it worthwhile and exciting. "People who are into this are passionate about it. I think because we intuitively understand that we need more nature in our cities. The figures are needed to convince someone who doesn't have the intuitive sense that this is worth doing". She has been pleased to find that the figures support her intuition. We talked about the overlap between areas which benefit from more nature in our cities and how much more could be done if all these areas contributed a little.

The recent increase in interest in green roofs and nature in the city Weeks sees as a continuance and progression of the environmental movement. Some interest, she thinks is driven by climate change and people's awareness of this. Weeks also consider that people want to be at the cutting edge and seen as leading the field. Information to facilitate this is shared more easily through media now. The Living Building Challenge has also progressed things. This is a certification where all different components are required. "It even has a beauty component. Of the standards that I know this is the most progressive and hardest to meet".

Weeks began describing her journey of passion that has brought her to where she is and how she has always been driven by altruism to help the environment. Her professional career has focused on scientific research that has helped provide the quantitative results to progress what she describes as her inherent knowing, her hunch. Not many people who I had met could so clearly articulate how their career was driven by their intuition and present a beautiful blending of the two.

Mary Davidge—17/12/2013
Principal,
Mary Davidge Associates
Green Building Consultant
Google HQ

Google has a reputation as being very "closed shop" and hard to visit or get much information from. The sense I had got in other conversations was one of frustration as people know Google is doing good things, particularly in design and biophilia, but they are not sharing their findings. I had heard from both Judith Heerwagen and Bill Browning that they were involved in consulting with Mary Davidge on design elements of biophilia. I was to meet with their contact at Google headquarters in Silicon Valley and get to join her for Google's famous buffet lunch.

I began with the obvious question: Why is Google interested in green buildings and biophilic design? "It's not just about productivity but one of Google's values is health and wellbeing of their employees. This is why we have worked with Terrapin Bright Green and others because we are looking for data about what aspects of the built environment can support human health and well-being. And not just being healthy, not just getting the toxic things out of the building but to create a more vibrant and energising user experience. And out of all the things we think we can do that is definitely one of the ones that is most promising".

She pointed out some immediate design features such as the daylit stairs and views to nature. "We know it's a better experience and it's going to affect their long term health". In Davidge's conversations with her employers, she is told that their charter "is to help Googlers live 30 years longer than the average person. So we need to think about what would really help that. What are the goals and what type of environment do we want to create? What is most promising? It's pretty clear that one of the things is access to nature. It's obvious. So the why is straightforward. The how is more difficult".

I commented on how this seemed to reflect a care Google has towards their employees and Davidge agreed, remarking how they wouldn't serve the quality of food they do if they were just concerned about productivity. Davidge has been working with Google for nearly 10 years and keeps seeing genuine concern for the health, well-being and experience of their employees. "So I think that's the why, that's simply put. Research tells us there is a psychological and physiological impact of having access to nature, so why not do it?"

We talked about the increase in creativity aspect which I had witnessed reflected throughout many of my interviews. Davidge commented that in talking with people many of them express that "they are more in that state of flow when they have access to nature. I think research proves this". I told her about the indoor biofilter living walls in Toronto and she told me that they did have some of these inside Google as indoor air quality was really important. Although the air was not so bad in Silicon Valley, Davidge explained that what they did here was for Google's headquarters worldwide. "Sometimes this necessitates only indoor biophilia".

Davidge explained that her company doesn't design but develops the guidelines for the designers. "We help the design team use the guidelines. So for instance on biophilia, if we were to say that they need to incorporate these four strategies we help them by doing case studies on where they would use these four strategies. What technologies they might use. We have a case history of what has worked for us in the past so each design team doesn't have to start from the beginning". I thought how much this made sense and that governments could take note. Davidge has to ensure that any project has enough biophilia and at least has achieved the baseline amount. Her role is to identify what are good ideas based on what has worked previously. I commented on what a fantastic case study and data set they must have here, especially as the results are global and they are compiling a lot of information. Davidge agreed. All data is owned by Google. Employee surveys are also carried out. I suggested that perhaps direct conversation and interviews with employees could gain more information. Davidge thought this was a good idea as the surveys do not reveal clearly what aspects of the biophilic design people are mainly responding to.

Google also has LEED strategies and her work helps in this as well. Google is just in the process of beginning some ground buildings which will incorporate more biophilic design. So far, the work has been to retrofit the existing buildings. "Going forward each project will have a requirement to meet those baseline biophilic metrics. I think it will grow over time" (Fig. 3.40).

The interview with Davidge was highly enjoyable combined with the pleasure and excitement of being at the famous Google Headquarters. Someone early on in Google made some good decisions as to the direction of the company and its employee relationships. It was very clear it was all about the health and well-being of the workers and encouraging that it was also very clear that biophilic design was the answer, especially when they have global researchers at their disposal. It is a shame that this amazing research and innovations are not shared publicly, but Davidge did suggest that one day in the right manner this may be possible.

Jeff Joslin—18/12/2013
Director of Current Planning
San Francisco Planning Department

Joslin had lived in Portland for 23 years and so knew Liptan and Houck very well, having also worked with them. Both he and Liptan recently went for a green roof tour in Switzerland which Joslin thought had been very "cataclysmic" for green roof efforts in San Francisco. Joslin trained as an architect and has always had a sustainability interest. He was involved with working with Hock and Liptan on the original bonuses and incentives that were part of Portland's planning code. "Hopefully I will be repeating that experience here".

Fig. 3.40 Google headquarters showing some of the grounds and Mary Davidge sharing a Google lunch (*Source* Author)

Joslin had been on the east coast of the US having lunch with a colleague after having just accepted his position in San Francisco. The colleague was the founding chair of Green Roofs for Healthy Cities (Stephen Peck) who informed Joslin that they were considering the next green roof conference being held in Seattle. Joslin suggested they have it in San Francisco instead, especially because San Francisco was so far behind in green roof initiatives. There are no incentives, requirements or specific programs. Joslin hoped that "the conference would jumpstart the conversation and start leading us towards some policy and regulatory elements". So, they did. He thinks it was very successful as previously there had been no real internal dialogue around the concept of green roofs. "We are not very practised at working together across departments and this was clearly the kind of technology and effort that was going to require that". One of the first things that happened after the conference was the formation of a San Francisco branch of SWISSNEX, a Swiss government initiative which helps educate other cities about green roofs. This is how the tour of green roofs in Switzerland had occurred. It included academics, industry representatives, policy advocates and government. "It was an incredibly informative trip, for me particularly, because we saw very directly, in the case of these cities, what happens as a direct result of the right policies and programs. They have been doing it in Switzerland longer. As Tom Liptan is the 'godfather' of green roofs over here, Stephan Brenneisan is the Swiss equivalent". Joslin explains that Switzerland has been building green roofs so consistently and for so long there is no longer a need to discuss the efficacy or reasonability. They are at a point where they are studying now how green roofs contribute to biodiversity.

Joslin personally feels that San Francisco does not need 3–5 years to figure out if green roofs work there or not. "I am more of the thinking if we build them they will come". One of the things he learnt in Switzerland is that sometimes they fail. Joslin clarifies that they might fail as a plant culture but all the other functionalities are there such as storm water remediation, filtration, insulating value, promoting roof life. The option is to refine the planting or wait and let nature take over. Joslin thinks that left to their own devices true believers, early adopters or others who are motivated will continue to add to the green roof inventory, but very slowly. He explains that at the point where you introduce incentives, regulations or a blend there are predictable results. First, he says, there is a steep incline in the proliferation of green roofs. Next there is a drop in the cost of the green roofs as the market responds to the increased demand with more products and increased expertise available.

Again the impact of the conference is reiterated as educating the San Francisco community of the potential benefits of green roofs. Joslin also mentions the document created by SPUR and the significance of this in aiding the creation of a road map for green roofs implementation in San Francisco. He feels that this document and the conference galvanised a high level of participation between primary agencies. These agencies are the Department of Planning, the Public Utilities Commission, the Department of Public Works and the San Francisco Office of the Environment.

Joslin hopes that ARUP and Weeks will keep bringing together the collective economic case as this is helpful in developmental conversation. He says it is there in bits and pieces but not collectively. When there is the economic case and avoided cost calculated, it is easier to translate this into financial incentives, Joslin explains. I asked Joslin if there was a major driver for green roofs in San Francisco. He explained that because of the multiple benefits there may not be the overarching driver but a number may contribute. I also asked him if there had been much conversation about human health and well-being of having increased nature in the city. "That's been more of an internal conversation than a public conversation, but I think we know intuitively it has a lot to contribute".

We discussed green walls and Joslin agrees it is harder to make the economic case for them, even though they are aesthetically pleasing and well-liked. I told him about the indoor green walls in Toronto that I was so impressed with. The local San Francisco green wall at the Drew School is not considered sustainable by his department, yet he does acknowledge the aesthetics and pleasure that people get from it.

Joslin then took me through his presentation for *CitiesAlive*. In it were density maps of San Francisco and we discussed the number of new developments being built. Most places I had visited in the US had exhibited signs of financial crisis, with building and development slowed. I asked Joslin where was their crisis? "There is no crisis here. But when you look at green roofs we are at the bottom of the heap". It was interesting he related these two points but we both tacitly understood why. In other cities, green roofs had frequently been in response to crisis. "There is fear concerning that we don't have everything we need in place. We don't have a tool kit in place and we haven't had a road map. And we haven't had Tom!" I commented how often these 'local champions', the risk-takers, have been vital to progression. Joslin agreed. "His story is very methodical. Tom saw roofs elsewhere. He thought intuitively it made sense. He stuck one on his garage. He monitored the shit out of it and came up with some results. He took them to his boss who thought it was interesting and decided to scale it up to see if it works on a public roof. All results were scalable and it went from there". In these few short sentences, Joslin had summed up two core repeated themes of my journey: the intuitive knowing and the local champion risk-takers.

Again Joslin presented a recurring theme and one that I related to as occurring in my home town. "San Francisco has a special challenge. We have silos. We have bureaucracies who don't talk to one another". Joslin showed me photos in his presentation of a green roof on a local building that had not been maintained and died after the building had changed owners. The green roof then naturally re-foliated and is now considered successful. San Francisco also has a green roof on their City Hall similar to Chicago. Joslin considers that as a result of the conference San Francisco has a number of champions and they have a road map, two components that were missing prior.

His presentation included relating the global challenges to San Francisco. "And we have the obvious of climate change and sea rise. Sea rise is very real to us here because we are always seeing it. It's not good, we are experiencing storm surges. For me it's not about sustainability, it's about survivability. We need to keep passionately making the case".

Joslin was a perfect last interview. He summed up the San Francisco story beautifully. So much of what he mentioned were the recurring themes of my journey which I think reflected the depth of both our understandings and learnings. I saw many similarities between San Francisco and my home city. There are possible impending crises, but neither city has yet faced the ones that other US cities are experiencing. Thus, the need for the local champions, who, as Joslin says, and I have heard repeated, have an "intuitive knowing". I could sense Joslin's desire to be prepared and to compile the tool kit. After his involvement with Portland, I also get the sense he is the person for the job. I enjoyed his attitude, which is similar to mine, in that the proof is there for the benefits of nature in the city. We don't need to reinvent the wheel in every city; we just need to do it.

California Academy of Sciences—20/12/2013
Kendra Hauser tour guide

I had the opportunity through my contacts to be taken on a private tour of the California Academy of Sciences, a science museum and a scientific and educational institution. The Academy has the largest green roof in San Francisco. We started with an interior tour and I discovered a wonderful green wall which had been created as a backdrop for the educational talks. The wall uses the same techniques as the Patric Blanc living walls and is created by a man who trained under Blanc. My tour guide, Hauser, says the response to the wall by the public is very positive. "People think the wall is more applicable to having at their house rather than having a living

Fig. 3.41 The educational green wall at the California Academy of Sciences (*Source* Author)

roof. People love a big green roof but I think they get a little daunted by the scale of it" (Fig. 3.41).

Hauser offered to tell me the story of the current building's design. When they were deciding to redo the building "the thinking was that a building was needed for the twenty-first century and one that represents our commitment to science and education. We have researchers all over the world. The Academy thought it was a natural movement to have science education and a green building because it is another way we can teach about science. It is another way we can teach about how to be sustainable, about how to preserve the natural world. It wasn't a push to make it happen, we didn't need to convince anyone. Everyone felt this was the next step".

The Academy is a double LEED platinum and very energy efficient, with recycling initiatives. The building has a lot of natural light, not only to save on lighting but also because the architect wanted the building to feel open and transparent so people could feel everything, and it creates a "piazza feel" in the building. The windows also act as heating and cooling ventilators. The natural light is also for the health of the employees (Fig. 3.42).

We then went upon the roof. It is a stunning roof with many curves. The architect described his vision at the tender stage, "It will be like a giant hand had lifted the park up and put a building under it". The curves sweep into seven domes which represent the seven hills of San Francisco. Within these domes are the windows, which not only give light to the public but also to the animals and plants housed below which provides another level of complexity to monitor and factor in. "The idea is also to encourage people to interact and become interested in green roofs, to encourage them

Fig. 3.42 Views of the interior of the dome roofs and greenery within (*Source* Author)

to go away and think about them further. We want to promote green roofs". I asked her why and she listed the reasons:

> They have functional benefits for the building such as increasing the roof's life span, heating and cooling savings, storm water management, sound reduction and for urban heat island reduction. They are also great for biodiversity and we can encourage wild life on our roof. We have species you don't normally see in an urban environment. The third aspect is for the human benefit and being able to look out onto greenery, so a health benefit. Green roofs provide amenities for the public plus the possibility of growing food. So we try to educate the public around all the benefits.

The biodiversity on the roof has increased from an initial 9 species to currently over 80 species. We try to encourage biodiversity by creating habitat niches such as logs and rocks and also a bee boxs (Figs. 3.43 and 3.44).

There are volunteers who come regularly to help maintain the roof. Hauser relates "how a lot of them come for the nature interaction as they may have indoor jobs, or for respite or to relax".

YouTube—20/12/2013

Again it was only through a contact that I was able to have a personal tour of the YouTube headquarters. The building has a green roof that undulates. There is no access, except for maintenance but it is visible through windows. They do water the roof a bit but basically leave it "to what grows grows". It helped achieve LEED platinum rating (Fig. 3.45).

Fig. 3.43 Kendra Hauser discussing biodiversity on the California Academy of Sciences green roof (*Source* Author)

Fig. 3.44 The insect home (*Source* Author)

Throughout the building, there were scattered biophilic design features. One was the use of fractal patterns in the creative think tank room. Images of flickering fires were on a tower of screen and the idea was to help the employees to step outside their normal environment and to step outside the box in their thinking. There are atriums for light and indoor trees in a mini forest arrangement (Fig. 3.46).

I met with a lot of people in San Francisco and saw a lot of initiatives. It was a perfect finale to my tour as many of the concepts that had arisen along my journey were presented in this city.

Everyone I interviewed, apart from Davidge at Google, mentioned the significance of the CitiesAlive conference. This can be an important stage as it unites the interested people. The barrier of siloed departments was recognised and mentioned frequently. Another interesting point was that there was no significant 'villain' as there had often been in other cities, there were potential ones. This means there was no motivating urban crisis. San Francisco, as Joslin mentioned, is behind other cities in green roofs. It was dependent on the local champions who were acting on their passion and intuitive knowing.

Fig. 3.45 The entrance and rooftop of YouTube (*Source* Author)

Fig. 3.46 The creative think tank and an indoor mini forest (*Source* Author)

Fremantle, Western Australia

Freo Greenskins was a collaborative project sparked by a particular event which occurred early in 2012. The City of Vincent, one of the Perth metropolitan councils, took the bold step of imposing the construction of a green wall as a building condition for a developer of a block of units. The new block was going to obstruct the view enjoyed by the tenants of an existing block of units, so the condition was the compromise. The developer took the council to court and won, with the condition being removed due to lack of precedence and a lack of knowledge about the practice which could be involved. This example had quite an impact on me as I thought if councils were going to take this bold step of initiating these building conditions then it was a shame to lose the opportunity for a green wall due to a lack of precedence and knowledge. This case was discussed between academics at Curtin University Sustainability Policy (CUSP) Institute and the Mayor of the City of Fremantle with the decision to trial a green wall in Fremantle so the precedent and knowledge was established. The evidence for the benefits of green walls was there, but the question of whether it would be successful in the challenging climate of Perth was unanswered

These two interviews are with the Mayor who approved the project and the contractor who received the tender to build the wall.

Julian Rose—07/05/2014
Director Deep Green Landscaping

Rose is the director of the landscaping company who was successful in gaining the tender for the Fremantle green walls. In the lead-up to the installation of the green walls he met with the trial's steering committee a number of times plus we had field trips to view his other constructions.

We met for an interview where Rose outlined the history of his business and the barriers he had encountered. He had been keen to build green roofs and green walls in Perth, Western Australia for some time and it had been slow, but consistent progress. Rose had been frustrated by some of the attitudes and lack of understanding that he encountered, but steadily persevered, meanwhile educating himself by attending conferences and researching.

At the time of the interview, his landscaping company, Deep Green, was working on the new hospital being built in Perth. The company had installed some green podiums and been involved with the building's landscaping. Rose was requested to cover a roof with blue metal, a common practice in building, but suggested putting a green roof there instead, as this roof has high visibility from surrounding buildings including future wards. From his research, Rose was aware that the benefits of patient recovery could be significant. The developers were in agreement, but the client, the Western Australian state government, was not prepared to pay the extra to have a green roof installed. Rose made the decision to pay the difference himself as he believed in "the multiple and significant benefits a green roof could make here".

Rose had initially been approaching developers in an effort to motivate them to include green walls and roofs in their constructions but had met with resistance, primarily because of the lack of precedence and justification for the costs involved. He then decided to approach local councils as he was aware that in many other countries these initiatives had been mandated and possibly the local government approach was the way to proceed: "In Germany there is a tax on storm water runoff which green roofs can help alleviate".

Approaching the councils had brought some success with some council funded projects. Currently, Deep Green is working with Belmont and Bayswater councils. When they had approached Subiaco council they had discovered that they were

Fig. 3.47 Subiaco library green wall (*Source* Author)

already interested and investigating the possibility of incorporating biophilic features within their districts. Deep Green was commissioned to build a green wall outside the Subiaco library. This has been very successful (Fig. 3.47).

When asked about the barriers he had encountered, Rose listed them:

- There are higher initial outlay costs and little understanding of the long-term benefits.
- Also Perth still lacks local precedence and knowledge about green walls and green roofs.
- Developers are unwilling to take the risk or what they perceive as risk.
- We also have the extra constraint of water conservation in Perth.

Rose did not think any of these barriers were insurmountable. It was just time and perseverance. He thinks that a point will be reached when it "finally catches on".

In a further conversation held in late 2014 regarding Fiona Stanley Hospital, Rose revealed that no green roofs went ahead on the hospital due to the roofing supplier refusing to honour his roof warranty if a green roof was installed on top.

I admired Rose's tenacity and faith that the market for biophilic design elements would eventually develop through education and demonstrations showing that it could work in a climate such as Perth. The barriers he was encountering were the same ones that I was experiencing in my immersive research journey. There was interest, yet a reluctance to spend the money and take the risk, especially when there was little precedence. Water issues are a very real concern in Perth and I was finding this was presenting as significant.

Brad Pettitt—21/07/2014
City of Fremantle Mayor
I met with Mayor Pettitt in July 2014 to discuss not just the Greenskins project but also the motivation behind the many other greening initiatives that were taking place within Fremantle, especially since the place-making workshop with David Engwicht in 2011. As with the Greenskins project, I was aware that Pettitt, with his team, had been responsible for 'driving' these.

When asked why he had been so active and committed to the increase of biophilic initiatives within the City of Fremantle Pettitt responded:

> I do think the future of cities is one where we are going to have real greenery such as green walls and green roofs contributing in terms of aesthetics, in terms of biodiversity and in terms of cooling microclimates plus the other sustainability benefits. But why I liked the Greenskins project was it was interesting in how we do this in a hot, dry place like WA, so I am hoping that this research contributes to making this possible. It is very different here to other places such as Singapore where research has been carried out. So this was the motivation, to demonstrate that we could do it and the lessons that are learnt.

Pettitt explained how the City of Fremantle has a Green Strategic Plan and a One Planet Living Strategy with a focus on creating a liveable, sustainable city. They are committed to planting one thousand trees a year in their greening policy. "Liveable cities are green cities". I asked him what his definition of liveable cites was and he gave me the following definition:

> Liveability of city centres is concerned with the quality of public spaces and the built environment. Liveable cities are ones in which people want to live, visit, spend time and linger – just like a nice room of a house.
>
> More formally, the concept of liveability has been linked to a range of factors, for example, quality of life, health, sense of safety, access to services, cost of living, comfortable living standards, mobility and transport, air quality and social participation.

Pettitt explains that it is easier greening from the ground up, implementing 'the low hanging fruit' with pop-ups and street pot plants. The walls are the hardest bit. He wants to work out what is viable to grow vertically and which species work or don't work. Pettitt finds people respond to the greenery, but to him this is about creating cooler, nicer, more liveable places. "People seek cooler, shadier places. People are very interested by the greenery and want it replicated on a bigger scale. It generates a lot of interest. It's happening all over the world so how can we do it in Fremantle?"

I asked him whether he thought that green roofs, green walls and other biophilic initiatives could be incorporated into policy, but Pettitt thinks we are a fair way from mandating any policy. Possibly green roofs could be mandated in larger scale developments. Pettitt suggests that where there has been policy implementation significant water management problems have driven this. I pointed out that in some places, such as Chicago, it has been a heat response.

This led to a discussion of what he perceived were barriers to implementation. "Barriers would be the cost, water use, maintenance, and that the fact that it's not yet mainstream so it's still being worked out". While his staff accept it and think that this is a good idea there is an awareness of the high cost and ongoing maintenance issues of green walls particularly.

Pettitt is waiting on the results from the Greenskins trial to ascertain whether things would progress further in this direction. These results are needed to determine whether green wall initiatives provide, as he says, the 'best bang for your bucks'. At the moment, street trees and small park pop-ups are more mainstream, accepted and an expected part of a liveable city.

3.5 The One Word and Connection to Nature

All interviewees were asked for one word to describe their feelings when they were immersed in nature. This provoked some reflective thinking for some while others knew immediately how they felt. Peace was the most mentioned word. The words are depicted in a word cloud in Fig. 3.48.

Implied within this is the idea that people recognise that they do feel different in nature. No one challenged the question; no one said that they feel the same as they would in purely built form urban density or anywhere. Peace could be said to be the antonym of stress, a common ill of contemporary urban lives. The implication is there that nature does offer what the literature suggests: a time to de-stress, to feel calm and connected and wonder at the beauty of it all.

Predominantly, interviewees had established a connection to nature when young. It was an even divide whether this occurred through urban nature, or rural and wilderness. It highlights the importance though, to create the opportunity for urbanites to establish a connection to nature through urban nature and not just beyond human settlement.

Fig. 3.48 Word cloud of collected words (*Source* Author)

3.6 List of Interviewees

See Table 3.1.

Table 3.1 List of interviewees

City or State	Interviewee	Word	Connection to nature
Virginia	**Tim Beatley** Teresa Heinz Professor of Sustainable Communities School of Architecture University of Virginia	Calm	Urban nature when young
Washington DC	**Michael Lucy** Senior Consultant Greening the Built Environment Anacostia Watershed Society	Home	Rural, young
	Kelliann Whitley Senior Property Manager Blake Real Estate	Peaceful	Urban nature, young
Pennsylvania	**Helena van Vliet** Biophilic Architect	Peace	Urban nature, young
New York	**Mary W. Rowe** Managing Director and Head Global Initiative for Urban Liveability and Resilience The Municipal Art Society of New York	Majesty	Rural, adult
	Bill Browning Director Terrapin Bright Green	Whole	Urban nature, young
	Stephen Kellert Tweedy Ordway Professor Emeritus Yale University School Forestry & Environmental Studies	Connectedness	Wilderness, youth
Chicago	**Jay Womack** WRD Environmental *Sustainable Results*	Inspired	Rural, young

(continued)

Table 3.1 (continued)

City or State	Interviewee	Word	Connection to nature
	Michael Berkshire Sustainable Development Division City of Chicago	Calm	Rural, young
Toronto	**Brad Bass** Great Lakes Issue Management and Reporting Section Environment Canada	Calm	Rural
	Roger Hansell University President at Noble International University Emeritus Professor, Department of Ecology and Evolutionary Biology University of Toronto, Canada	Peace	Rural
	Birgit Siber Principal Diamond Schmitt Architect	Connected	Rural, young
	Steven Peck Founder and President Green Roofs for Healthy Cities	Connected	Wilderness, young
Vancouver	**Randy Sharp** Principal Sharpe and Diamond Landscape Architects	Relaxed	Rural, young
	Cornelia Hahn Oberlander Landscape Architect	Relief	Urban nature, young
Seattle	**Dave LaClergue** Urban Designer Department of Planning and Development I City of Seattle	Interested	Urban nature, young
	Judith Heerwagen HPGB Program Expert Office of Federal High-Performance Green Buildings US General Services Administration	Calming	Wilderness, young

(continued)

Table 3.1 (continued)

City or State	Interviewee	Word	Connection to nature
Portland	**Matt Burlin** Sustainable Stormwater Management Environmental services City of Portland	Peace	Wilderness, adult
	Mike Houck Director Urban Greenspaces Institute		Rural and urban nature, young
	Tom Liptan Ecoroof Program Manager at City of Portland Portland, Oregon Area Environmental Services	Wonder	Urban nature, young
	Ada Kardos Portland homeowner	Contentment	Urban nature, adult
San Francisco	**Kirstin Weeks** Energy and Building Ecology Specialist ARUP	Vitality	Urban nature, youth
	Mary Davidge Principal, Mary Davidge Associates Green Building Consultant Google HQ		
	Scott Edmondson Planner/Economist Information and Analysis Group San Francisco Planning Department	Wonder	Urban nature, young
	Peter Brastow Senior Biodiversity Coordinator SF Environment	Peace	Rural, young
	Jeff Joslin Director of Current Planning San Francisco Planning Department	Humble	Urban nature, adult

(continued)

Table 3.1 (continued)

City or State	Interviewee	Word	Connection to nature
Perth, Australia	**Mayor Brad Pettitt** Mayor of the City of Fremantle	Peaceful	Urban nature, young
	Julian Rose Director Deep Green Corporation		

3.7 Conclusion

The journey through the US, Canada and my home town was extremely rewarding. People were forthcoming and generous with their time and with their knowledge. The journey resulted in meeting significant players in biophilic design who had made substantial contributions to the social movement.

In many cities, the literature was being translated into practice. There was a focus on green roofs as the major biophilic design element, mostly implemented for their environmental benefits. The expressed social benefits reflected the research literature. Many of the interviewees were aware of the multiple benefits of biophilic design or had discovered them after implementing a design feature. Interviewees shared an appreciation of and connection to nature, sometimes expressed as an emotion or love of nature. For many, the initiatives they were implementing were common sense in creating more liveable cities and in increasing the human–nature connection.

The interviews, and subsequent writing of the stories, identified many motivators and drivers in both the emergent and coalescent stages of the social movement. These were compiled into a matrix for each interviewee which were then amalgamated, creating huge lists of overlapping social, environmental and economic motivators and drivers in each social movement stage. Other themes that recurred throughout the writing of the journey were noted.

Chapter 4 contains the analysis of the journey and discussion of the motivators, drivers and revealed components that contributed to the social movement of biophilic design.

References

Beatley T (2011) Biophilic cities: integrating nature into urban design and planning. Island Press, Washington, DC

Biederman I, Vessel EA (2006) Perceptual pleasure and the brain: A novel theory explains why the brain craves information and seeks it through the senses. Am Sci 94(3):247–253. https://doi.org/10.2307/27858773

Cramer JS, Browning WD (2008) Transforming building practices through biophilic design. In: Kellert SR, Heerwagen J, Mador M (eds) Biophilic design: The theory, science, and practice of bringing buildings to life. Hoboken, Wiley, New Jersey

Friends of the Highline (2013) Highline field guide. Friends of the Highline, New York
Kellert SR (2008a) Dimensions, elements, and attributes of biophilic design. In: Kellert SR, Heerwagen J, Mador ML (eds) Biophilic design. Wiley, New Jersey, US, pp 3–19
Kellert SR (2008b) Biophilia. In: Sven EJ, Brian F (eds) Encyclopedia of ecology. Academic Press, Oxford, pp 462–466
Kellert SR, Wilson EO (eds) (1993) The biophilia hypothesis. Island Press, Washington, D.C.

Chapter 4
How the Biophilic Design Social Movement Informs Planning and Policy

Abstract A new social movement called *biophilic design* has emerged over recent decades to enable greater connection between nature and urban populations. Understanding the origins of this social movement can inform further progression, implementation and planning of biophilic design. This chapter's research methodology of immersion and heuristic inquiry within the social movement recognises that there are three stages of a social movement. In each stage, there are particular motivators and drivers that provide impetus to the movement as revealed in interviews with the pioneers of biophilic design. These are translated into a framework of ten actions for mainstreaming implementation of biophilic design attributes in urban environments.

Keywords Social movement · Biophilic design · Nature · Cities · Biophilic urbanism · Biophilic drivers · Biophilic policy · Implementing biophilic design

Research Highlights

- Social movements progress through three stages: emergence, coalescence and mainstream.
- Biophilic design emerged as a social movement primarily driven by five motivators.
- Successful trials, multiple benefits and strategic policy helped the coalescence.
- Learning from these motivators and drivers informs a framework for mainstreaming.
- Multi-sector collaboration including government is needed for delivery structures.

4.1 Introduction

Biophilic design suggests there is an innate human need to affiliate with nature in the built environment. Although a relatively recent concept, it appears to have become a social movement, defined as a

> loosely organized but sustained campaign in support of a social goal…they result from the more or less spontaneous coming together of people whose relationships are not defined by rules and procedures but who merely share a common outlook on society.
>
> (Smelser n.d.)

J. Söderlund, *The Emergence of Biophilic Design*, Cities and Nature,
https://doi.org/10.1007/978-3-030-29813-5_4

This chapter examines how the biophilic design movement has happened and what can be learned from this by other cities and places wanting to have a similar process.

Social movements begin with a core group of people uniting with a shared social goal, to enable some kind of change (Jasper 2007). They are categorised into three stages of development: emergence, coalescence and mainstream (Christiansen 2009). In order to understand biophilic design as a social movement, an immersive journey of heuristic inquiry was undertaken. The chapter will examine how this was done and the results that were found.

These understandings of what has driven and motivated the progression and implementation of biophilic design, contribute to creating a framework of actions for mainstreaming biophilic design. This framework can be utilised by governments and organisations, to inform strategic planning and development within cities.

4.2 Research Methodology

Chapter 3 presented exemplary individual stories of North American pioneers in the social movement of biophilic design.

The research involved two Australian and twenty-eight North American interviews, in ten cities, of the leaders, the forerunners and the local champions of biophilic design in their city. The interviews utilised a heuristic inquiry framework of semi-structured conversations that were structured sufficiently to obtain core information but allowed to 'ramble' to a natural completion. This encouraged passions and emotions to be revealed and a deeper exchange between the interviewer and the interviewee to take place. Each conversation or interview resulted in a tabular analysis at the end with a snapshot of the extracted components. This analysis is based on two key principles of doing this kind of qualitative research:

1. The extraction of the emergence motivators, coalescence drivers and mainstreaming policy implementation was based on word usage and responses to direct questioning. Literature research provided a list of rationales motivating or hindering biophilic design initiatives plus keywords associated with these. Where these occurred in an interview, it was noted and they were listed in the interview matrix. Reflection on the written interviews revealed any other words which were frequently used in discussion around motivators, drivers and barriers. At the conclusion of the interviews, the phrases and words which compiled the matrices were listed and a word frequency analytical tool used to reveal the common motivators, drivers and barriers. The motivators and drivers were then themed.
2. In all qualitative research, especially with conversations, there is a certain amount of bias as the process relies on the researcher's perceptions and the focus that has been chosen to interpret the data. The interviews were recorded as well as extensive notes taken so there is accuracy in what was spoken and the words

which were used. As a result, there are word usage patterns that emerge from all the interviews, and these have been used to frame the analysis.

The compiled matrices from the Australian and global interviews contained comments which expressed the motivators as experienced by the interviewee and interpreted by the researcher. These comments were themed and are visually expressed in the mind map, Fig. 4.1. Three representative comments were chosen for each theme and are depicted on the smaller branches which lead to the themes comprising the emergent and the coalescent stages of the social movement. The lighter lines depict the links between the emergent theme and the coalescent theme, that is, where there was a flow on from one stage to another as revealed in the interviews.

Reflection on the research notes and the interviews uncovered shared thoughts and observations on the dynamics, the flows and the interplays of biophilic design initiatives. These are also discussed. Further to this, the influencing components of mainstreaming include a discussion of the barriers as expressed by the interviewees plus discussion of other significant themes that emerged as the basis for mainstreaming. The explicated themes were utilised in determining a framework of principles and practices that would progress the social movement of biophilic design from an emergent stage to a coalescent stage and into an accepted, mainstreamed stage.

4.3 Prime Motivating Themes in the Emergent Stage

Kellert, in his interview, considers there is not a prime motivator to designing biophilically: "The multiple benefits that can ensue from biophilic building provide a list of potential motivators, each of them important" (Kellert, personal communication, 2013).

The prime motivator themes revealed in the emergent stage are the core internal emotions that trigger the initial stirrings, the themes that unite or lead the decision-making towards biophilic design. It may be what initially motivates an individual on a course of action before the awareness of a greater social movement.

4.3.1 Connection to Nature

Beginning with the initial group who discussed the biophilia hypothesis, the desire to increase the human connection to nature in cities has been evident as a prime emergent motivator. This tends to be coupled with a recognition of the profound disconnect to nature that is currently the dominant design paradigm of cities. Interviewees recognised that this lack of opportunity to connect with nature is a recent phenomenon which is impacting on the health and well-being of urban children, contributing to the obesity crisis. The 'free range' child is disappearing and possibly,

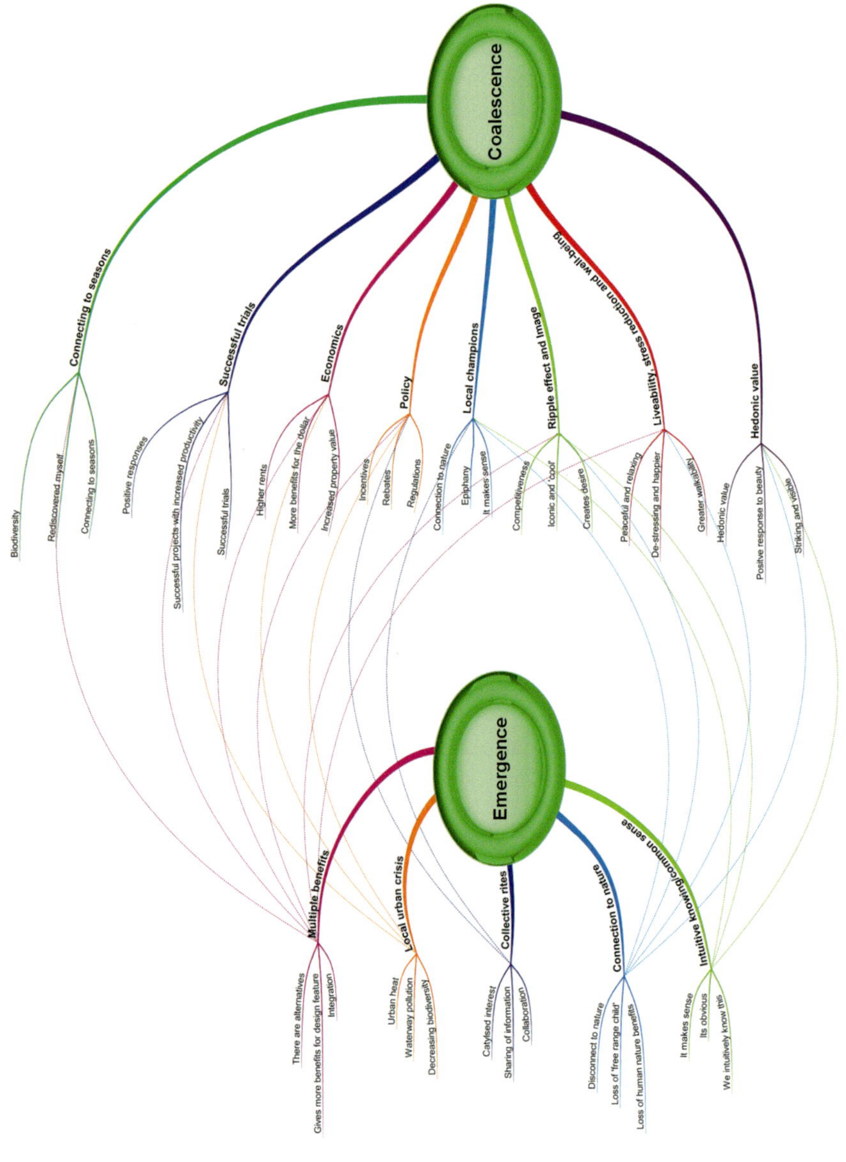

Fig. 4.1 Composite depiction of the motivating themes in the biophilic design social movement (*Source* Author)

alongside this, great environmentalists whose childhood involved significant contact with nature.

Other interviewees expressed a desire to increase people's connection to nature that was also driven by the hope of increasing local inhabitants' connection to place and thus their stewardship of the area and the flora and fauna

4.3.2 *Intuitive Knowing/Common Sense*

A large number of the interviewees expressed the view that biophilic design was 'intuitively obvious', 'common sense' or it 'made sense'. This was particularly apparent when the interviewee had investigated design options that fulfilled their criteria and became aware of a biophilic option. The view that it was common sense was strengthened if there was awareness of the multiple benefits on offer with a biophilic approach.

Other interviewees framed this differently, on a deeper level. This understanding appeared to be what motivated them more on an emotional level than intellectual. They mentioned that it was intuitively obvious or there was an intuitive knowing. This is indicative of an inner resonance for some people with the concept of biophilic design.

4.3.3 *Local Urban Crisis*

A local crisis is perceived by some and not others so can be called an internal motivator that drives change. The crisis triggers a response. The need to address a local crisis has been a major motivator especially in the government arena. Many of the North American cities have water pollution crises created by the CSS (Combined Sewer System) water management system, or from large amounts of storm water run-off from impervious urban surfaces. With the US Federal government mandating the necessity of dealing with this issue, as well as concern from communities, many biophilic initiatives such as green roofs and rain gardens have emerged as a result.

Urban heat featured as the other main local crisis which has driven a biophilic design response. This initially occurred in Chicago in response to heat-related deaths, but also featured in other cities, such as San Francisco, which have awareness of heat issues even if there have not been significant deaths as in Chicago heatwaves.

Less drastic crises, but still significant, are decreasing biodiversity, climate change and sea level rise. San Francisco, a city with less rainfall than the other visited cities, is primarily focused on these concerns. Poor air quality, especially indoors, has prompted technological innovations in Toronto. For Portland (and similarly Seattle), it is management of water pollution and river degradation which was threatening their salmon industry, combined with the need for increasing density due to their urban growth boundary. There was recognition by many of the interviewees that biophilic design can contribute to doing density better and retaining a place for nature within the built environment. In the ten US cities visited, biophilic initiatives in response to an environmental crisis were a strong and predominant motivator which also enabled the necessary supporting framework such as research and policy to be enacted with some speed.

4.3.4 Multiple Benefits

Among the interviewees, there was recognition that there were alternatives to biophilic initiatives in achieving similar outcomes particularly in response to an urban crisis. Awareness of the multiple benefits that can come with biophilic design had frequently been a deciding factor in pursuing the appropriate biophilic design element. Of note though, is that the awareness is siloed within the sustainability arena where the interviewee and the prime motivator sit. For example, when implementing green roofs for storm water management, there was awareness of the multiple benefits of potential energy consumption reduction, reduction in urban heat and increased biodiversity but little discussion of the possible social and economic benefits. Likewise, if an interviewee's area and interest was in the social benefits of biophilic design, there was a tendency to focus on the multiple social benefits and not so much the environmental or economic.

4.3.5 The Collective Rites of Conferences

The emergence of biophilic design involved two conferences, or meetings; where first biophilia was discussed, and second, where the biophilia hypothesis was discussed. Without these initial gatherings, the social movement of biophilic design may not have erupted. These collective rites of bringing like thinkers together are essential both in driving the movement and in motivating the people who attend them.

Conferences featured strongly throughout the interviews as having provided a forum for groups and departments to come together and begin a dialogue: a collective rite. Frequently this also provided the jumpstart a city needed through the education of participants and by galvanising a high level of participation between primary agencies. Conferences had spurred the development of 'local champions'. For example, the sharing of information through a collective rite has enabled a city and its inhabitants to have the confidence to proceed in the implementation of green roofs in their city.

Conferences and collective rites provide the opportunity and space for the arenas of civic society, government, industry and academia to come together. This is an important space as from these collaborative meetings, direct action and implementation of biophilic design may occur.

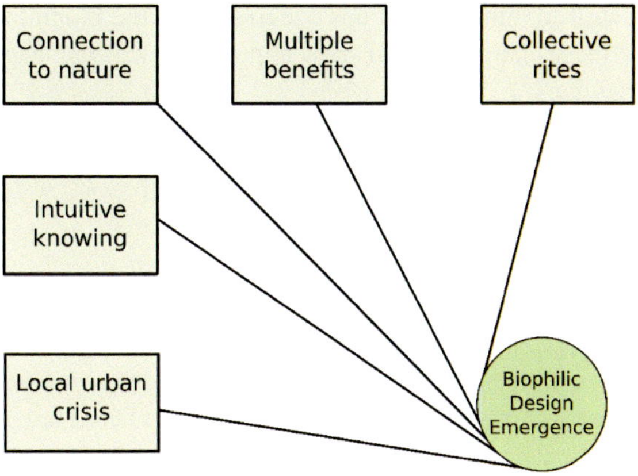

Fig. 4.2 Prime motivating themes in the emergent stage (*Source* Author)

The discussed prime motivating themes discussed above are depicted in Fig. 4.2.

4.4 Prime Motivating Themes in the Coalescent Stage

In the coalescent stage of social movement, leaders have emerged, barriers have begun to be overcome through successful outcomes and exposure, clear strategies have emerged and the movement is gaining momentum. This stage reveals the drivers to start change.

4.4.1 Local Champions

Throughout the interviews, the realisation grew stronger that many of the interviewees were leaders and local heroes in their field. They trialled new techniques, sometimes at home, and were innovative in their ideas. At times they had to stand strong against resistance from industry, government or even community. Some of these local heroes were wise elders who, with the experience and wisdom of their years, had a deep understanding of the benefits that nature and biophilia could bring into people's lives. They had experienced it and witnessed it. They were great leaders who had sometimes extended beyond their boundaries with the knowledge and beliefs that were a lifetime's accumulation. Some of the interviewees knew each other and referred to the other as a hero or 'local champion'.

Most of the interviewees could be classed as 'local champions'. The influence, and necessity, of local champions was dramatised repeatedly in the interviews and people's stories. Many of the cities owed their biophilic initiatives to the passion and belief of one or more local inhabitants. Most of the interviewees had come to an understanding that cities and nature can, and need, to integrate and coexist. This is a shift in thinking that has occurred with the biophilic design movement. For some of these champions, this new approach can attract negativity from other community members.

4.4.2 Successful Trials

Most cities seek demonstrations before proceeding further with any new idea. The progression of implementation of biophilic initiatives would not have happened if the early trials were not successful. This was repeated throughout the cities. The success of the green roof trial on Chicago City Hall gave a precedent and ability to ask the same of the developers. Chicago has had success with many of their policies, but not all, which enables them to recognise and build on what works.

The high level of successful research in Toronto has made the city a leader in green roof innovations. While most of the measured outcomes have been in the environmental area, successful social and economic outcomes are also major drivers for biophilic design. Both Chicago and Portland had stories of developers who had experienced unexpected economic benefits through installing a green roof and attracting higher rents and keen tenants.

4.4.3 Multiple Benefits

Multiple benefits are both an emergent driver and a coalescent driver. The difference is that as a coalescent driver it is more about the discovered benefits after installation. Many of the related stories, particularly around green roof implementation, were about how they were initially constructed for environmental outcomes. Other benefits, the social, economic and unplanned environmental, were 'discovered' after the installation. Often mentioned were the peace, enjoyment and de-stressing benefits that came with easy, and every day, access to nature.

Observation and experience of the biophilic design initiative, as related by the interviewees, consistently revealed further benefits beyond the initial targeted benefit. This translates to a greater understanding of all the environmental, economic and social benefits that are associated with biophilic design elements. A significant number of interviewees commented on the exposure of the multiple benefits inspiring them further.

4.4.4 Strategic Policy

The success of trials, particularly with green roofs, was a vital stepping stone towards enabling the introduction of strategic policy in some cities. In Chicago, the trial occurred within the government arena with a successful green roof on City Hall. This alone was not enough to progress the adoption of green roof initiatives by developers which is why strategic policies were created to progress green roofs in Chicago. Green roof standards were included in Chicago's sustainable development policy. After initial resistance, developers discovered they enjoyed positive responses and further benefits from clientele. City of Chicago's own research, as related by the city planner, revealed that 82% of green roofs in Chicago were installed because of policy.

In Portland, Oregon, legislation promoted the use of green roofs in response to the federal *Clean Water Act*. This was after successful trials by local champions who took their successful outcomes to a receptive member of government. It was a similar story in Toronto.

Biophilic design suitably sits under the umbrella of a variety of policies. By adopting a general policy statement, or by aligning with the goals of the global Biophilic City Network, cities can work towards encouraging increased uptake of biophilic design initiatives.

4.4.5 Ripple Effect and Image

A consistent observation from interviewees was the experience of the ripple effect. This is where observation of a feature leads to the observer wanting one also and acting on this desire.

Demand for biophilic initiatives can be driven by the consumer base who, by witnessing examples, request similar from developers. It can ripple through the design and construction industry if it is seen to give a competitive advantage by this increased demand. This is how a market demand is started for biophilic urbanism in this coalescent stage though all the other factors combine to provide its ongoing market demand as a more mainstream phenomenon.

4.4.6 Aesthetics

People respond to beauty. There is proposition in the research literature that human responses to aesthetics are prompted by evolutionary survival instincts and are a component of innate biophilic tendencies (Kellert 2008a, b; Heerwagen and Gregory 2008). The academic literature has shown that humans have positive psychological responses to aesthetics and the varied textures, patterns and motions of nature contribute to people's appreciation of beauty. There are often alternatives to biophilic design that are less attractive. People choose the biophilic approach as it contributes to the aesthetics of an area, so it brings another benefit beyond the function. Building design based solely on function and technology to meet sustainability standards may not be attractive for people, which may make it unsustainable. The aesthetics of biophilic design addresses the human factor of sustainable design.

People are willing to pay for views of nature, which demonstrates that there is a hedonic value attributed to an aesthetically pleasing building.

Landscape architecture is a well-established profession and practice, typically brought in at the end of a building project to 'beautify' the landscape. Biophilic urbanism would suggest that landscape architects need to be employed earlier in developments than established customs, and in a more wholistic manner otherwise the deeper aesthetics are lost as token plantings are made. Developers, according to the interviewees in each town, are becoming more interested in creating a stronger link between their developments and biophilic design because people seem to respond to it.

4.4.7 Economics

The economic case for biophilic design did not initially drive the first demonstrations in the cities under study. The first steps were seen to be an extra cost but after a while, this seems to have merged into a much broader acceptance that the economics of biophilic design was manageable and that other benefits included potential to be an economic success story. Owners and developers of buildings which incorporate biophilic design elements were found to have discovered economic rewards particularly through higher rents and increased property value. In commercial buildings, biophilic design led to retail sales increasing and hospitals enjoying the economic benefits of shorter hospital stays and less stress among patients and visitors, especially in waiting rooms. Most interviewees had stories of demonstrations of sometimes unexpected economic benefits which had inspired further biophilic design projects. An increasing recognition of the multiple benefits that result from biophilic design emerging was that many interviewees suggested there was growing recognition that there is good return for the money spent.

4.4.8 Liveability, Stress Reduction and Well-Being

Some of the interviewees had personally discovered how access to a green roof was relaxing and peaceful and contributed to their well-being. Other interviewees related stories of behaviour they had observed and anecdotes they had heard which demonstrated people's responses and appreciation of increased access to nature in the urban environment.

The many qualities of biophilic design contribute to increased well-being and liveable cities. Engaging people with nature through gardens or shaded streets increases the walkability of a city and the health of its inhabitants. Stress reduction and happier,

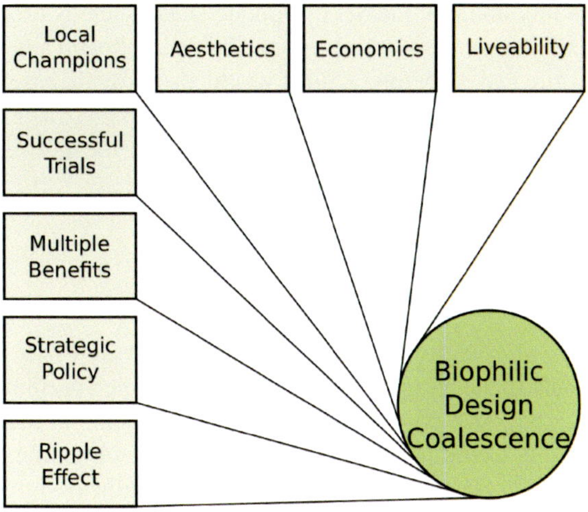

Fig. 4.3 Prime motivating themes in the coalescent stage (*Source* Author)

healthier people were frequently mentioned outcomes which inspired further design initiatives.

The eight prime motivating themes in the coalescent social movement stage of biophilic design are illustrated in Fig. 4.3.

4.5 Prime Motivating Themes for a Mainstreaming Framework

The biophilic design social movement was largely at the coalescent stage when the interviews were collected. Few people suggested that it was mainstream yet, though many felt it could become so eventually. Derived from the immersive journey, the interviews and the composite depiction set out above, the components of mainstreaming the biophilic design social movement can be formulated.

4.5.1 Identify the Need

Where there were clear urban crises which biophilic design could assist in mitigating, research, trials, local champions and policy had a defined pathway to a desired outcome. Not many urban crises are unique to a city; rather they are shared globally but may vary in their intensity. Each city needs to identify their driving issues as a unifying starting point.

A city inventory or 'stocktake' would assist in helping to identify the current state of the city. Surveys of the urban environmental condition which include biodiversity quality and risk factors of climate stress such as urban heat and flooding could be undertaken. A social inventory of people's health and well-being plus their access to nature and daily interactions would also provide base indicators from which policy-makers, city planners and designers could plan strategies towards a healthy, biophilic city. This assessment may also help provide the motivation and rationale for biophilic initiatives.

4.5.2 *Identify and Address Barriers*

Many of the barriers to biophilic design which present in a city are not unique to that city but have been encountered and overcome in the forerunner cities of biophilic design. Common barriers which this research journey identified are outlined below and, as can be seen, are not unique to biophilic design but can have global relevance:

- Higher initial costs,
- Ongoing maintenance costs,
- Siloed benefits,
- Lack of understanding,
- Lack of precedent,
- Fear of new technologies and associated risks,
- High water use,
- Lack of leadership,
- Lack of political will and
- Financial mechanisms not reflecting the benefits, especially in health.

Some barriers are perceived and tend to arise from lack of understanding and knowledge and so can be addressed by education and learning from other cities experience. This is where conferences, workshops and joining global organisations, such as Biophilic Cities, can be useful.

Following this framework for mainstreaming actions for implementation, gives hope to these barriers being overcome.

4.5.3 *Unite Silos Through Collaboration*

Silos and separation between professions appear to be common within societal structures globally. Biophilic design is interdisciplinary, combining design professions, academics, engineers, developers, planners, health professionals and others. Facilitating interdisciplinary communication on biophilic design data and research may help streamline implementation, increase efficiency, lead to further innovations and improve responses to urban issues.

Government, industry and community are also important, and involved, stakeholders. Conferences and symposia provide the opportunity to bring together these disciplines and stakeholders. Collaborative meetings such as these addresses the barrier of siloing and exposes the potential for interdisciplinary multiple social, environmental and economic benefits. Being aware of the interdisciplinary nature of biophilic design and the multiple benefits that can ensue through implementation also lays the groundwork for financial contributions from stakeholders. An example is the environmental benefit of cooling cities through direct nature and greenery. This reduces urban heat and heat-related health stresses and deaths (such as the previously discussed Chicago response). Health costs for treating heat-related issues would diminish. 'Should Health Departments then contribute towards greening cities?' is a question which could be discussed.

The need for biophilic designers to be collaborative is obvious. This is one profession and indeed one social movement where you can say that nearly every profession and area of knowledge needs to be involved in the process of delivering biophilic design outcomes. Collaboration also contributes towards the development of best practices in biophilic design, synergistic benefits and the opportunity to involve, support and further educate self-organising efforts from civil society and thus build natural capital.

4.5.4 Enable Local Champions

From these collaborative gatherings, or driven by a personal interest and understanding, the potential for the emergence of local champions exists. These are the people to whom the wholistic framework makes sense and who are prepared to step up and out and push through any barriers. The way they do this is usually through telling stories.

The importance and power of stories were mentioned by the interviewees, and many told a story to demonstrate a concept and an outcome in a powerful manner. People relate to these and in an emerging movement such as biophilic design, stories can reach many people in different arenas, beyond the academics and professionals. They are easily understood, providing real-life examples and proof. Stories can connect people to nature more, encouraging them towards greater stewardship of their environment and beyond. They can teach things about nature in a way with which people can identify. Many of these stories have found their way into books (e.g., Beatley 2011; Matan and Newman 2015; Newman et al. 2017; Söderlund 2016).

The collected stories from interviewees in the research journey, especially the local champions, revealed more information and increased understanding beyond academic research. The stories revealed the motivators of either the city under discussion or the personal ones. They also illustrated both social movement theory and biophilic concepts. Beyond that though, the stories helped to characterise the collective identity of the biophilic design social movement.

4.5.5 Create Partnerships

The stories collected throughout the research journey are stories of pioneers and innovators in biophilic design, mostly working alone to begin with. Many then formed small groups in order to get underway. Once they had a sense of being in possession of a solution, all of the key players found partnerships where they could begin delivering the practical outcomes.

Inspiring local champions and fostering collaboration will potentially lead to the creation of partnerships with the idea of delivering biophilic design to the market place. The challenge will be for governments at all levels to cultivate these collaborative partnerships, particularly between industry, investors and researchers, and recognise the importance of supporting emerging industries and new technologies. These partnerships can support each other by creating, or improving, the culture for innovative thinking where barriers can be removed, investment increased, there is greater diversification and the market for innovation works better.

4.5.6 Set-up Demonstrations

Collaboration, local champions and partnerships supported by, and including government, will enable more, and better, case studies and demonstrations to occur. The benefits of these initial implementations of biophilic design principles are multiple:

- They provide the meeting place for the different players and arenas.
- They provide the precedence and can break through barriers and rationalise risks.
- They offer the opportunity for research, data collection and innovation.
- Implementation of biophilic design elements can lead to post-completion understanding of the full range of benefits through experience and observation.
- Observation of the initiative could stimulate market demand.

The Chicago City Hall green roof is considered iconic. Its importance is due to its early implementation as one of the first major green roofs in North America, as well as the fact it was installed as a case study and demonstration. It was designed to study thermal properties, plant survival rates, storm water retention and, importantly, to educate. The roof was successful in all its goals and green roofs proceeded to be implemented throughout Chicago. It established a precedent. The Chicago City planner could show developers who were reluctant to comply with policy how it was done and how it looked. Demonstrations and case studies are vital in providing the tangible showpiece and knowledge to motivate further implementation.

Demonstrations are not only important in setting precedents and providing knowledge, they also provide a visual example which could trigger the ripple effect of implementation as experienced in some cities. Some developers, investors and clients such as the Washington DC embassies have green features installed as they enjoy being an early adopter, seeing an opportunity to be unique and iconic. For others, it may be the observed social amenity and aesthetics that motivate them to instal their own. Bartczak et al.'s (2013) research into people's decision-making processes towards installing a living green wall concluded that the wall's contribution to building aesthetics was the most important factor in the process. This, possibly hedonic, value of green infrastructure contributes to the ripple effect of copycat installations that can follow from the initial demonstration.

4.5.7 Integrate into Professions

The research journey reinforced in every city what the literature had already revealed: a lack of integration among the professions. This is not unexpected as biophilic design is still not a mainstream educational opportunity nor are there professional organisations that have set up to provide the integrated science, psychology and design elements of biophilic design. The professionals interviewed had largely developed their expertise through their own experience and through demonstration projects. Mainstreaming will need new educational programs and new professional bodies to be formed. Biophilic design will be mainstreamed when it is integrated into urban design, landscape architecture and architecture as an everyday practice and not seen as an add-on. The first stages will require academics from backgrounds in environmental science, engineering, humanistic psychology and design at all scales, to form integrated teaching programs that can bring out the core principles and practices of biophilic design. When mainstreamed, biophilic design will be a standard part of education and will be prolific throughout the literature.

Professional practice will need to be modified through demonstrations that can help to rewrite the manuals of built environment practice. This will be a process of trial and error with rapid learning that is passed on throughout the growing social movement across the globe.

4.5.8 Create Delivery Structures

Conferences

Conferences are a particular practice of society that has powerful delivery outcomes. People who come probably already share a common identity, however loosely, but at conferences, stories can be shared and exchanged, themes can be explored and unified plans made. Face-to-face conferences provide a forum for the aligning of frames and recognition of the common identity. In five of the cities visited, conferences had been held which had an impact on the progression of biophilic design. They brought together players from different arenas with a common problem needing to be solved. These conferences can also provide the framework and education to foster local champions.

Collective rites at a conference can bring together people who otherwise may not meet. They may be from very separate arenas yet have a common goal, or common problem to address. Uniting these groups through collective rites is essential for establishing a functional wholistic framework for urban design. Indeed, the initial group whom Kellert brought together were from diverse backgrounds, yet as one interviewed attendee mentioned, they focused on creating more pleasing, happier and healthier cities, with the diversity adding to the richness of the outcome.

Government

Policy has been successful, particularly in the implementation of green roofs in global cities, by incorporating the option for a green roof in sustainability standards, or by providing incentives and rebates, and by regulation. Germany began with incentives in 1983 followed by Basel, Switzerland, in 2000. Both countries currently have a high number of green roofs, and it is now an accepted form of practice. Since then, Chicago, US; Tokyo, Japan; Portland, US; Toronto, Canada; Seattle, US; and most recently France have all initiated some form of incentives or regulations encouraging green roof construction in their cities.

Chicago successfully implemented policy which led to a substantial increase in the number of green roofs in the city. Chicago City found that regulations, the stick, worked better than incentives, the carrot. But there was a difficult time in the initial implementation due to a backlash from developers and planners which needed the mayor and the city planner to stand firm.

Globally, there is a growing tool kit of options for policy.

Civic

Community desire may also present as a policy driver. The power of the ripple effect and aesthetic, hedonic responses could lead to demand for biophilic design being led by community and the supporting biophilic design industry. The rippling effect of biophilic design initiatives supports the research literature and the biophilic theory of an innate and responsive connection with nature that, once triggered, creates a desire for a greater connection. This also suggests that the divide from nature in cities has not been a divide of humans and nature but has been driven by politics and economics as well as modernistic architectural design that cut out nature from buildings.

Government recognition of community impetus in biophilic activism is crucial for enabling policies that support, not interrupt, the community and any positive self-organising initiatives. The government can also play a role in providing literature and an interface between the research and the civic community.

Industry and business

New planning and design practices need to synthesise across the interdisciplinary professions and community to build biophilic design into cities. Biophilic values and goals need to be incorporated in city planning, design and development plus be included in strategic economic development. Sustainability or 'green' ratings for buildings should include a biophilic component. This integration of biophilia and ecosystem services needs to occur at all levels, from region to city to precinct to block to building.

4.5.9 Make the Business Case

In the interviews, the strongest barrier to the implementation of biophilic design in cities was the upfront costs. Even where the costs were comparable but different design thinking was needed, interviewees expressed an encountered resistance to varying from business as usual.

Presenting research, perhaps through a conference, outlining the social and environmental benefits and how these translate to economic benefits would help to provide the business case for both government and industry investors. Detailing technical know-how and product information would assist to alleviate the perceived risks associated with the lack of knowledge and understanding. If the multiple social, environmental and economic benefits were understood in all arenas, it would assist in uniting the arenas towards a combined integrated approach. This would hopefully provide greater means and funding for biophilic design implementation.

In addition, while a complete cost-benefit analysis which includes financial assessment of the multiple benefits would be challenging in its scope, much research could be done within defined parameters that presents a solid business case. For example, life cycle analysis could strengthen the case for green roof implementation by factoring in the savings from the extended roof life, reduced energy consumption and water management plus increased biodiversity and the health and social amenity benefits. Comparative life cycle costing of non-biophilic alternatives may also strengthen the case for biophilic options.

Ideally, to shift away from economic analysis as a main driver would be the most beneficial. If, for example, a green roof was installed solely for the economic savings in energy consumption and did not deliver the expected financial outcome, then further green roof projects could be stalled. Viewing a city as a functioning natural system where the benefits of biophilic design were understood to address a variety of the city's issues would strengthen the business case by removing the emphasis on economics as the sole criteria. The human economy depends on the natural economy: on biodiversity and functioning ecosystems. This interdependence needs to be factored into policy decision-making and a biophilic design approach can provide an interface between the two.

4.5.10 Link to Academic Research

In almost all the cases examined and the people interviewed, there was a link to academic research. The social movement needs its work validated and enabled through academic literature. The divisions in academic research between professional disciplines mean that biophilic design is slower to flourish. While the research benefits are there in the literature under their niche headings, their lack of integration necessitates an interpreter: someone who will do the exploratory research into the literature and champion the cause. The next phase of mainstreaming will need more integrative

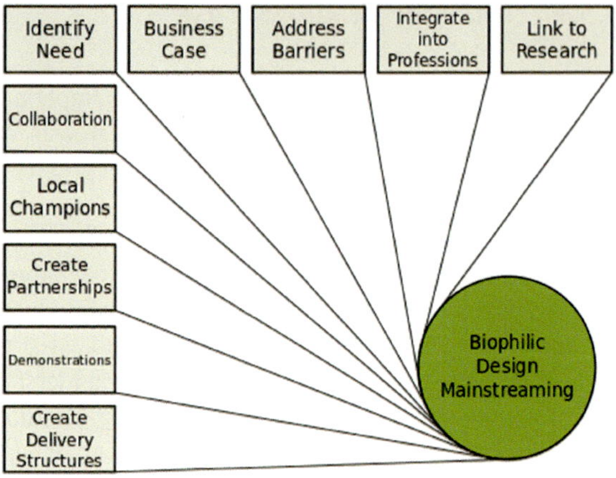

Fig. 4.4 Prime motivating themes for a mainstreaming framework (*Source* Author)

research and the people involved in the social movement will need to seek out links to academic research.

The framework of actions for mainstreaming the social movement of biophilic design are diagrammatically expressed above in Fig. 4.4.

4.6 Connection and Creativity

A more subtle but repeated theme that emerged from interviews throughout the immersive journey in varying terms was the intuitive, deeper sense of the dynamics of the human connection with nature. It felt like an underlying jigsaw tugging to be put together. Connecting to nature and its beauty draws humans in, captivating them through innate biophilic responses, triggering psychological and physiological reactions which restore and de-stress people. While this response may vary with individuals, for some it also taps an intuitive knowing: a sense of wonder and a sense of peace. With the body and mind relaxed and receptive, what some interviewees call an 'open mind', creativity can flow. Ideas and innovations can be born, answers can be found.

With increasing urban growth, there is acknowledgement among thinkers that answers to modern-day issues are to be found in cities. Cities are a dynamic mix of function and aesthetics, an organised complexity that, like nature, has the potential to self-organise; especially when threatened or facing crises. Immersion, stories and conversation illustrated the varying depths of responses people have with exposure to nature. Providing greater opportunity for human exposure to the positive benefits of nature in the built environment enhances the capacity for positive feedback loops into the self-organising dynamic of the civic arena. These positive feedbacks may contain technological innovations which contribute to restoring urban balance as a functioning ecosystem; and they may contain innovations and technological advancements which enable further implementation of nature's patterns and spaces. This would increase the opportunity for urbanites to connect with nature, and the cycle begins again.

Enhancing the opportunity for the nurturing of biophilic tendencies with the resulting positive innate psychological and physiological responses not only helps individuals to manage everyday life in a city, it also assists to build social 'nature fitness'. This 'nature fitness' could also facilitate social connectivity. As discussed, the research literature has shown people are less stressed in biophilic spaces and more likely to be creative. This can enhance social creative discussion and connection. The success of community gardens illustrates the appeal of social interactions around nature.

To build cities devoid of nature may contribute to the impairment of the self-organising system to innovate, adapt and increase its resilience. Government policy can facilitate the process, or it can hinder. If policy is transparent and collaborative there can be a meeting place for top-down and bottom-up enterprises. Within this junction, community ideas and innovations can be valued and implemented. Collective rites such as conferences can facilitate the collaborative process between government and self-organising community.

4.7 Summary

The main requirements for mainstreaming biophilic design emerged as a continuous thread of integration, collaboration and unification. Where the arenas of civic society, industry, government or academia had met there had been movement and outcomes in the biophilic design social movement and biophilic urban initiatives. The separation of social and environmental benefits in the research was reflected in the explicated themes, presenting as a barrier due to its contribution to the segmenting of the benefits in practice. Uniting the research would assist in uniting the arenas and the stakeholders plus reinforcing the economic business case with a more wholistic representation.

Collective rites, collaboration, local champions, case studies and demonstrations can unite and integrate, with connection with nature and creativity providing the interface for government and community, the natural and social economies and all the biophilic benefits to overlap in a synergistic self-organising urban culture response to a need to survive. Biophilic tendencies can be nurtured and strengthened.

Biophilic design is providing a framework for this integration and the term potentially can unite a diversity of disciplines, benefits and design elements. Experience and growth has revealed the multitude of aspects and elements that can come under the term biophilic design, perhaps more than the early group who met and discussed the concept ever thought. There is a rapidly growing shift in the approach to how cities are being designed and a term for this such as biophilic design is helpful yet there is the possibility that biophilic design as a term may be outgrown or refined. Figure 4.5 illustrates the motivators and drivers progressing towards mainstreaming biophilic design.

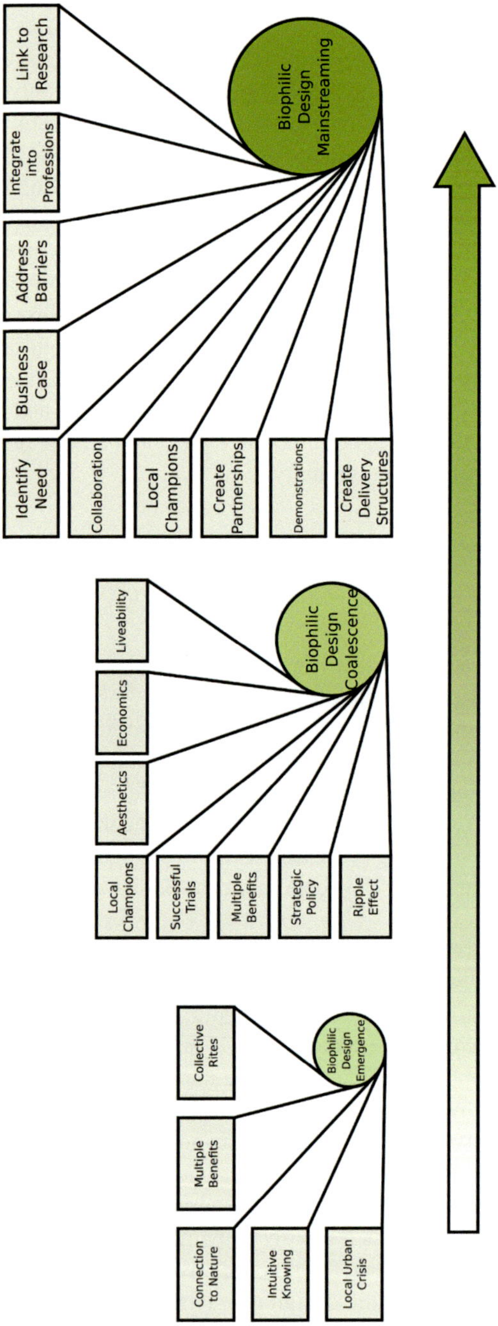

Fig. 4.5 Biophilic design social movement journey (*Source* Author)

4.8 Conclusion

A love of life, or biophilic connection to nature, is a consistent theme that has driven the social movement of biophilic design. People in cities need greater connection to nature in their daily life. The biophilic design social movement appears to have emerged and even coalesced in some of the cities studied. However, for it to mainstream, as happens in successful social movements, it will need to develop a series of components that were identified and suggested by the local champions in each city. The research has shown a social movement with similar motivators, drivers, themes and components based on the integration of biological/environmental science, psychology and design. The interviewees told a story and created a vision of possibilities where biophilic design is an integrated component of city design. This assisted in the development of a framework for mainstreaming biophilic design initiatives in global cities with ten actions.

The social movement of biophilic design has triggered a shift in the approach to the inclusion of nature within cities. The world is ready and, it is possibly willing, to mainstream biophilic design but it will take significantly more of the identified actions to make this happen.

References

Bartczak C, Dunbar B, Bohren L (2013) Incorporating biophilic design through living walls: the decision-making process. In: Henn R, Hoffman A (eds) Constructing green: the social structures of sustainability. Massachusetts Institute of Technology, Massachusetts

Beatley T (2011) Biophilic cities: integrating nature into urban design and planning. Island Press, Washington, DC

Christiansen J (2009) Four stages of social movements. Research Starters; Academic Topic Overviews, EBSCO, Ipswich, MA

Heerwagen J, Gregory B (2008) Biophilia and sensory aesthetics. In: Kellert SR, Heerwagen JH, Mador ML (eds) Biophilic design: the theory, science, and practice of bringing buildings to life. Wiley, New Jersey, pp 227–241

Jasper J (2007) Social movements. In: Ritzer G (ed) The Blackwell encyclopedia of sociology. Blackwell, Malden, MA

Kellert SR (2008a) Dimensions, elements, and attributes of biophilic design. In: Kellert SR, Heerwagen J, Mador ML (eds) Biophilic design. Wiley, New Jersey, US, pp 3–19

Kellert SR (2008b) Biophilia. In: Sven EJ, Brian F (eds) Encyclopedia of ecology. Academic Press, Oxford, pp 462–466

Matan A, Newman P (2015) Green urbanism in the Indian Ocean region. J Indian Ocean RegN 11(1):60–73. https://doi.org/10.1080/19480881.2015.1017228

Newman P, Beatley T, Boyer H (2017) Resilient cities. Island Press/Center for Resource Economics

Smelser NJ (n.d.) Social movement. http://www.britannica.com/EBchecked/topic/551335/socialmovement. Accessed 09 Sept 2014

Söderlund J (2016) Piloting green walls and studying their effects. In: Beatley (ed) Handbook of biophilic city planning and design. Island Press, Washington, DC, pp 156–157

Chapter 5
Living Green Wall Trials in a Hot Dry City Climate

Abstract Research on living green walls highlighted the potential for urban heat reduction, increased biodiversity and social benefits. A trial of living green walls was undertaken in two challenging (both socially and environmentally) inner-city locations. Environmental data was collected plus a social survey of peoples' responses was conducted. The survey revealed a willingness and desire to have more nature in the city. Appreciation for the beauty of nature received a slightly higher positive response than stress reduction or increasing happiness. Due to correct installation and species selection the living walls thrived. It was a collaborative project between a university and a local government with unanticipated benefits encountered throughout the project.

Keywords Living wall · Green wall biodiversity · Living green wall · Urban heat reduction · Stress reduction · Living wall trials

5.1 Introduction

The city of Perth, with the port city of Fremantle, had always been perceived as a big country town, isolated from the rest of Australia and the world. Its development had been slow and steady and, with lots of space, the focus was on the suburban family blocks with big backyards. With beautiful beaches, these suburban developments sprawled along the coast. The city had never experienced any major crisis, socially or environmentally.

This changed in the late 1970s with a mining boom. The state of Western Australia, where Perth is situated, has long been known for its rich mineral deposits, especially iron ore. In 1981, in response to market demand, the state experienced a mining boom (Ye 2008; Battellino 2010). There was a rapid development of mine sites and workers flooded in from around the world with stories of employment. The mining companies decided that it made financial sense to fly workers in and out of the mine sites rather than build towns and amenities in the area. Perth experienced a rapid growth in demand for housing; the demand was from companies and individuals experiencing the wealth that came with mining and exporting a desired commodity. The population increased by 400,000 in 7 years and passed two million in 2015.

J. Söderlund, *The Emergence of Biophilic Design*, Cities and Nature,
https://doi.org/10.1007/978-3-030-29813-5_5

Perth began to experience an affordable housing crisis (Worthington 2012). Rental properties in the more popular suburbs became expensive and in short supply. New housing was sought and there was rapid building in response to the demand. Land was cleared and Perth expanded outward, sprawling out to 120 km along the coast (Moran and Novak 2009). Inner-city apartments were also in demand, and the inner city experienced an unprecedented growth in density. Perth's typical low density (Curtis 2008) was beginning to change in response to demand as well as the necessity to curb the outward expansion with its associated bush clearing. New housing and precinct designs are now emerging that aim to create denser, more liveable suburbs. Australians typically love their outdoor living, gardens and backyards and hence were looking for a way to bring nature into the new denser housing options. Incorporating the elements of biophilic design may assist the embracing of higher density living.

Perth has a hot, dry climate with a decreasing rainfall. A sufficient water supply for the future is not ensured (Water Corporation 2009) and the city now has more than half of its water supply from desalination. Thus, the inclusion of direct nature needs to be water wise and utilise innovative water-saving technologies. The use of water though has the potential to help reduce urban temperatures by evapotranspiration and the creation of microclimates. The city's design focus has not considered the high and sustained temperatures of the summers. In many streets, there is little shade and pedestrians scuttle in the summer from shade spot to shade spot. Urban heat, though not yet a disaster, has the potential to be fatal if excessive heatwaves occur, such as Chicago, and recently India, have experienced.

Currently, there has been no urgent need for change through a crisis which has occurred; there is a need for change to avert the crisis which through global examples and predictions appears inevitable. Perth is fortunate to be able to learn through other cities.

Biophilic design initiatives offer the potential to cool the city and to facilitate high density living through aesthetically increasing well-being. This design approach, though, needs someone to start making it happen and convince others to invest in the design approach, be the risk-taker, the early adopter, the leader. Fortunately, the Mayor of the City of Fremantle was receptive to trialling green wall in his city.

5.2 Fremantle (Freo) Greenskins

Freo Greenskins was a collaborative project sparked by a particular event which occurred early in 2012. The City of Vincent, one of the Perth metropolitan councils, took the bold step of imposing the construction of a green wall as a building condition for a developer of a block of units. The new block was going to obstruct the view enjoyed by the tenants of an existing block of units, so the condition was the compromise. The developer took the council to court and won, with the condition being removed due to lack of precedence and a lack of knowledge about the skills and methods which could be involved.

This case was discussed between academics at Curtin University Sustainability Policy (CUSP) Institute and the Mayor of the City of Fremantle with the decision to trial a green wall in Fremantle so the precedent and knowledge were established. The evidence for the benefits of green walls was there, but the question of whether it would be successful in the challenging climate of Perth was unanswered.

5.2.1 The Project

A steering group was established comprising two academics from CUSP, the Mayor and the City of Fremantle's Park and Gardens manager. The group met regularly to discuss planning and criteria.

There were six major stages throughout the project:

- Identifying the site,
- Creating the tender brief,
- Calling for tenders,
- Designing the monitoring framework,
- Installation and opening event,
- Monitoring and
- Analysing results.

Identifying the site

Fremantle, having its origins as the port for Perth, has many heritage buildings from the late 1800s. It took some time to find a site that met with the criteria the steering group thought important.

- It was in the public eye.
- It was a council-owned building or the owners would agree to the wall's installation.
- There was an accessible water supply.
- The position would adequately test the climatic conditions that other green walls may have to: strong, salty sea breezes, radiant pavement heat and the north sun.

The Chosen Sites

Two challenging sites, both socially and climatically, were intentionally chosen. Both are in the Fremantle CBD. Site A is a north-facing wall in a little-used mall with significant antisocial behaviour. The area is a sun trap in summer with a high level of radiant heat from the paving and surrounding walls. Site B is a west-facing street scape wall, again in an area where antisocial behaviour occurs. It is exposed to the late afternoon westerly sun and strong sea breezes (Figs. 5.1 and 5.2).

The building on which the sites are located is privately owned and the owners were apprehensive and concerned about possible water damage. With the Mayor involved in the conversation and with reassurances, they gave their approval.

Creating the tender brief

Many variables and parameters needed to be clarified and defined such as fixing/support structures, maintenance plan, water sources and does each wall contain

Fig. 5.1 Site A (*Source* Author)

Fig. 5.2 Site B (*Source* Author)

similar species with similar water requirements? Are the species mixed to trial greater variety? Are species selected and swatches designed with potential water availability from sustainable sources as a parameter? Measurable outcomes were to include water use, plant performance and comparison of ambient, wall and canopy temperatures.

Call for tenders

Expressions of interest to tender had already been received by five local contractors. Green wall construction was a small and highly competitive business in Perth. This was a high profile job and there was not a lot of work in building green walls in Perth.

The contractors were keenly aware that their logo would be on the signage and their public exposure would be significant.

Once parameters had been clearly defined, a design criterion was created and sent to the five contractors with a 2-week submission deadline. Outlined within the criterion to be addressed were the following points:

- Design (including a schedule of all material types, images of example installation and details of irrigation system,
- Construction (including construction details of all joinery and wall attachment),
- Installation (including access and equipment requirements),
- Watering requirements (including plant species ET parameters and irrigation schedule),
- Maintenance (including frequency, tasks and cost),
- Species (differentiating between local native and exotic),
- Size of plants at time of planting,
- Cost (pro bono components to be identified if included) and
- Time frame (from award of contract to completion of installation).

Five tenders were received and the steering committee members assembled and selected the winning contractor. The contractor would construct and plant the green wall off-site, installing it after 3 months when there had been sufficient growth of the plants. This also meant that installation would occur in autumn, when climatic conditions are less harsh, and the plants would have time to settle before the test of summer.

5.2.2 *Designing the Monitoring Framework*

It was in this 3-month time period that the parameters and design of the social survey needed to be defined and the survey constructed. It was decided to link the survey to a QR code that could be placed on signage beside the walls and provide a direct link to the online survey.

Social survey goals

The purpose of the survey was to gain an indication of public response to building integrated vegetation in the urban environment. If their response was favourable, it was to then gauge what the prime motive for the positive response is. What draws them to respond to the survey? There is awareness that a strong negative reaction could still motivate them to fill in the survey. The survey and the green wall trial were also to gain some insight into whether there was any knowledge or understanding of the social and environmental benefits of nature. Would the responses reflect awareness of the physiological and psychological benefits of biophilic design? The questions explored several dimensions of response:

- Aesthetic: did people like how the walls looked?

- Design and function: did people feel the walls were useful and enhanced public space?
- Environmental: did the walls contribute to environmental amenity or urban biodiversity?

Respondents were also asked about their connection to nature and how much time they currently spent in nature or more natural surroundings.

Underpinning this were several questions:

- How do you assess the depth of the reaction?
- What emotions do the walls evoke? (framing indifference as an emotion).
- What aspect of the wall are they are responding to, e.g. the beauty, the 'green', the 'yah this might do something to this area', the environmental aspect?
- How interested are they in the concept and project?
- What are the emotions that are driving the interest?
- Would they support more green walls being installed and how strongly would they support this? What would they give up/ranking?
- How important is seeing more green in their urban environment to them?
- Would they like this in their home?
- Would they see it as saving the world? Are they interested?
- Are their feelings towards nature in cities influenced by whether they grew up in nature or an urban city environment?

These are big questions behind a small project but they were sparking the biophilic design social movement so considered essential.

Environmental monitoring goals

Having researched what had been monitored in other green wall trials such as Hort Park in Singapore, the decision was made to monitor the following parameters:

- Water usage,
- Temperatures in canopy,
- Temperatures behind wall,
- Temperatures of nearby blank control wall and
- Temperatures 15 cm out from canopy.

Eighteen temperature-only data loggers and six temperature and humidity data loggers were installed. Water meters at Site A would measure water in and water out, so evapotranspiration rates could be extrapolated. Plant growth rates would be regularly visually assessed to determine which species are thriving. The City of Fremantle would maintain and manage the environmental monitoring plus undertake the analysis at the conclusion of the project.

Fig. 5.3 Installing the wall (*Source* Author)

5.2.3 *Installation and Opening Event*

The installing of the green walls 03/04/13

The day of installation was very exciting. It carried the sense of a project coming to fruition. The green walls were planted with species chosen for their resilience in hot dry conditions, comprising a combination of local plant species and exotics. Both sites utilise fertigation methods to provide nutrients and have a monitored watering system (Fig. 5.3).

It was a long process to get them mounted on the wall and hooked up to the watering system. I visited the site at various times throughout the day to see how it was going and speak with the installers. By the middle of the day, the installers reported that many people had viewed the walls, with about 15–20 people stopping to inquire further. They said, "there was definitely significant interest". When asked what type of feedback they had received the answer was "If they weren't drunk it was positive!" This comment was indicative of the social issues that the area can experience. While I was there an elderly man stopped to talk. He was willing to complete a survey and so was the first to do so. Shortly afterwards, some young women stopped and spent quite some time there looking at the green walls, inquiring how they were done and wishing they could have one in their own houses. Both were happy to complete a survey. These were the first public responses to the installed green walls and the interest was encouraging (Fig. 5.4).

The Official Opening 12/04/13

A crowd of nearly 50 turned out to celebrate the installation of the two trial green walls in Fremantle. This was surprising considering that it was 9 am in the morning,

Fig. 5.4 Initial public interest (*Source* Author)

the only time convenient for the VIP attendees. A welcome to country by Noongar elder offered the support and good wishes of the traditional owners. The Fremantle Mayor briefly outlined the 18-month process of collaboration between the City of Fremantle and CUSP that resulted in the installation (Fig. 5.5).

The location of the opening was by one of the green walls in Westgate Mall. It is on a bleak wall in a little-used urban mall and much of the feedback from the attendees was extremely positive, with many recognising the social benefits that greenery can bring to otherwise underused places. The feel among the crowd was

Fig. 5.5 The opening day crowd (*Source* Author)

Fig. 5.6 Signage with QR code (*Source* Author)

one of inspiration, with talk of creating similar projects and many people wanting bigger and more of this type of living wall installation.

The CUSP director introduced the online survey developed to research the social response to the green walls and most attendees completed the survey at the event (Fig. 5.6).

The day after the opening a Federal member of parliament launched an initiative at the mall site. He spoke highly of the green walls, inviting me also to speak. Already the site was attracting more people. This was to be the trend.

5.3 The Response

As the project grew so did community awareness of the trial. With the installation and launching of the online survey, the interest rapidly increased. Our institute had an online blog with an article on the project, the survey and my email address. Regular emails were received from interested people. The queries ranged: how do I build a green wall or green roof? What are the regulations regarding them? Did I want to build a demonstration one on their roof? Who should they contact to get one built? What species should they plant at home? Did I need help?

Queries also came from media, with requests for interviews or information. There were articles in the local Fremantle newspaper about the project, plus a small one in the *The West Australian.* Discussions regarding the Freo Greenskins project were appearing in online blogs. The Biophilic Cities website posted an article about the project. They invited me to write a blog for their website and this was posted also. A short piece on the project will be included in two forthcoming books: one on resilient cities and another on sustainable development goals. A popular and well-respected Saturday morning radio show on gardening requested a live interview about the green wall trials (Lush and Hahn 2013). 202020, a national network of organisations campaigning for increased urban green (20% more by 2020), invited the Fremantle Greenskins project to be included (202020, 2014).

A significant request came from Curtin University. Twice a year they publish their 'flagship' magazine with two featured research projects being undertaken at the university. The Fremantle Greenskins project was chosen to be featured. This entailed photographs and an interview. The Curtin University website also showcased the project on their homepage for a week and posted the story in their news and events (Daniel 2012). When asked why the interest in this particular project, the answer was because it was perceived that "there is a lot of community interest in green walls and green roofs at the moment" (Daniel, personal communication, 2013).

In 2013, shortly after the Fremantle Greenskins opening, a representative from the Australian Federal government Major Cities Unit in the Department of Infrastructure and Transport approached me for an article on the project to be included in the 2013 report *State of Australian cities.* This signified that the interest was national and government was considering that biophilic design and urban greening was growing within our cities.

The amount and calibre of media, community and industry interest in some quite small green walls in Fremantle strongly indicated that the green walls were triggering something significant within our society. The social survey attached to the trial could provide an indicator of what that 'something' may be.

Although the green walls are situated in an area that can attract social problems, the walls have never been vandalised. The water monitoring system that is situated a few metres away did get tampered with, which negated our water use assessment. Small events have been held in the area. The area also attracted a graffiti artist that significantly improved the area (see Fig. 5.7).

5.4 The Social Survey Results

The response to the walls from the community and media had been very positive, but biased as it obviously was limited to people who were already interested enough to make contact. The survey would reveal whether similar attitudes were widespread in the greater community. There would still be a bias as the responses were only from those who cared or who were interested enough to take the time to complete the survey.

Fig. 5.7 Activity in the area (*Source* Author)

Survey respondents
Over a 2-year period from April 2013 to April 2015, survey responses were received from 83 people, though not all respondents chose to answer all questions. Of those who responded, 61 (74%) had actually seen the walls, while the remaining 22 people (27%) had not. People who had not seen the walls were not required to answer specific questions about the Fremantle Greenskins walls and asked only to respond to more general questions about green walls and connection to nature.

The sample size (n =>83) is not sufficient for inferential statistical analysis. The responses collected were sufficient to generate descriptive outcomes, though in most cases the sample size within demographic categories was too small to enable identification of significant difference. Even so, the data is useful in giving an indication of community response to green walls. In terms of age and gender, this sample could be considered generally representative of the adult public living in Fremantle.

Slightly more than half of respondents were female (56%) and slightly less than half were aged between 25 and 44 years (46%). Approximately half of all respondents reported living in Fremantle or surrounding suburbs.

The majority of people grew up in a city (60%) or in both city and country (26%). The majority of respondents reported spending as much time as possible in nature (32%) or regularly spending time in nature (50%) as a child.

Responses to survey questions

The survey consisted of 31 questions, including the demographic questions shown in Table 5.1, with several questions where respondents were asked to provide comment or a word to describe their response to the green walls. Written responses were analysed using thematic classification (Bryman 2008).

Two main question formats were used. One employed the semantic differential which utilises a scale where people are asked to place their responses between two polar adjectives, or extremes such as useful to useless. The other format used a Likert scale where people were asked to agree or disagree with a series of statements (Bryman 2008).

Questions and responses were grouped into several themes to simplify analysis:

- Urban design and function,

Table 5.1 Demographic profile and time spent in nature

	n	%
Gender		
Female	**43**	***56***
Male	34	*44*
Total	77	*100*
Age		
Under 18 years	2	*3*
18–24	9	*11*
25–34	**20**	***25***
35–44	**16**	***20***
45–54	14	*18*
55–64	10	*13*
65–74	7	*9*
75 or older	1	*1*
Total	79	*100*
Where did you grow up?		
City	**48**	***60***
Country	12	*15*
Both city and country	21	*26*
Total	81	*100*
As a child, did you spend time in nature?		
Yes, as often as possible	**26**	***32***
Yes, quite regularly	**40**	***50***
Yes, but not often	12	*15*
Yes, but rarely	1	*1*
No	2	2
Total	81	*100*

- Aesthetic (visual) response,
- Affective (emotional) response,
- Nature in the city,
- Connection to nature and
- Time spent in nature.

Responses to each of the survey themes are explored, with qualitative analysis of comments provided to round out the discussion.

There is a strong positive response to all questions (Table 5.2). The majority of respondents indicated they thought green walls were useful, and make a city more attractive and liveable, and help to reduce heat. An overwhelming majority of people (95%) responded positively to the suggestion that City of Fremantle should provide funds to construct more green walls.

Table 5.2 Urban design and function

The Fremantle green walls are…	**Useful**				**Useless**	
	60%	**30%**	**7%**	**2%**	**2%**	
	35	*17*	*4*	*1*	*1*	*58*
	Strongly agree	***Agree***	***Disagree***	***Strongly disagree***	***Don't know***	
Green walls can help make a city more attractive and liveable	**86%**	**14%**	**0.00%**	**1%**	**0.00%**	
	65	*10*	*0*	*1*	*0*	*76*
Green walls should be a standard part of urban design and architecture practice	**74%**	**20%**	**5%**	**1%**	**0.0%**	
	56	*15*	*4*	*1*	*0*	*76*
Green walls could help reduce heat reflected off streets and buildings	**63%**	**29%**	**1%**	**0.0%**	**7%**	
	48	*22*	*1*	*0*	*5*	*76*
Should the City of Fremantle make provision in their budget for the construction of green walls?	**Yes**	**No**				
	95	**5**				
	71	*4*				*75*

Comments included:

> More need to be implemented, as they assist in removal of air pollution, which will be increasing with climate change. They also make the area look great.
>
> It's a great idea that deserves to be expanded throughout urban areas and residential developments should be encouraged to follow suit.
>
> As an early intervention they are most useful in the green deserts of the CBD to inspire us to do more as a community.
>
> More green walls will introduce the concept of reducing heat island effect into the future to deal with a warming city and a greater diversity of indigenous plants may well entice indigenous birds back into the city.

In response to the question about future investment by the City of Fremantle, most comments were positive:

> Global warming and the psychological benefits ... make green walls an essential for budgets.
>
> Absolutely! Especially in spaces like this mall which is unused and sad to see in a beautiful place like Fremantle.
>
> I see this as a positive point of difference that will bring revenue to the city, along with all the benefits already discussed in this survey.
>
> Should be part of the built fabric in all future projects and all existing buildings should include green walls and roofs.

Some respondents indicated they would support more walls as long as they were bigger, and in more public places:

> But only if the size of the walls is significant enough to show benefits to people and the environment.
>
> As long as they're in a prominent position and are of a decent size. Like 10 times the current experiment.

Others queried who should pay:

> The question in my mind is [whether] this is a City of Freo expense or property owners.
>
> In general I believe the property owner or occupier who benefits from their installation should provide funding. So when I say yes, I mean yes for Council owned buildings or encouragement/incentives for owners to install.
>
> And local businesses that have the space should be encouraged to have them.

One respondent saw investment in green walls as a means to contribute to environmental and social amenity:

> I think there is merit in a contribution scheme to encourage owner installation. This could be through direct contributions, council commitment to maintenance or even providing information on installers. Also could be an interesting project by combining amenity benefits of living walls and food production for homeless/less fortunate as some kind of social scheme. I noticed the one that was outside the Council building in Kings Square installed by Sustainable Outdoors had some vegetables.

And some respondents felt planting trees would be a better initiative while another wanted to know why green walls might be better than planting trees:

Table 5.3 Aesthetic (visual) response to the green walls

The Fremantle green walls...	**Enhance the area**				**Detract from the area**	
	83%	**12%**	**5%**	**0.0%**	**0.0%**	
	47	*7*	*3*	*0*	*0*	*57*
The Fremantle green walls are...	**Beautiful**				**Ugly**	
	63%	**30%**	**4%**	**3%**	**0.0%**	
	36	*17*	*2*	*2*	*0*	*57*

No they should find any clear space on the verge and plant a combination of native and (fast-growing) non-native species. They should also encourage people (if they don't already) to plant native plants (small bushes etc.) on the council verge.

Given the tiny amount of carbon sequestration from a green wall people shouldn't bother – large, shady trees do a much better job of reducing the heat load in a city – green walls would use a huge amount of water during the summer (watered at least once a day).

Spell out the basics to Joe and Mary citizen – why do it? How is it better than just planting another tree? (i.e. to those of us who know nothing about green walls except that they look pretty, address the so what question – in language that we understand.

Survey results indicate that most people (83%) strongly agreed that the walls enhanced the areas they were located. There was positive response regarding their beauty—most people agreed (30%) or strongly agreed (63%) that the walls were beautiful (Table 5.3).

One respondent commented:

City of Fremantle should construct more green walls. The location for this one (Westgate Mall) is unexpected! But it enhances the aesthetics of this dilapidated mall, and creates an area of interest.

Many other comments described the walls as attractive and a welcome alternative to blank walls. Comments included:

It's nice to have greenery climbing the wall and brighten an otherwise dull area.

Changes for bleak, blank walls

Always lovely to come across … and so beautiful at present. Amazed at how healthy the wall garden is.

This last comment was received in August 2014 at the end of winter. A more recent comment regarding how the wall looked during summer was not as complimentary:

The one I saw (opposite Coles) was so high it was only by chance that I looked up and noticed it. That was January 2015 and given the relatively long hot summer it was looking pretty shabby – plus it covered such a small area as to not make an impact. Nice idea, but the climate is clearly not suited.

Survey results indicate that most people felt inspiration (63%) and wonder (62%) when looking at the walls. A small proportion of respondents reported feeling indifferent (5%) (Table 5.4).

Table 5.4 Affective (emotional) response to the green walls

The Fremantle green walls are...	**Desirable**				**Worthless**	
	80%	**10%**	**7%**	**2%**	**1%**	
	46	*6*	*4*	*1*	*1*	*58*
When I look at the walls, I feel...	**Inspired**				**Indifferent**	
	63%	**22%**	**10%**	**0.0%**	**5%**	
	36	*13*	*6*	*0*	*3*	*58*
The Fremantle green walls are...	**Wonderful**				**Pointless**	
	62%	**28%**	**9%**	**0.0%**	**1%**	
	36	*16*	*5*	*0*	*1*	*58*

Feelings of inspiration and wonder, and observations that the walls positively affected how people felt, were reflected in several comments:

> Let's do more for Fremantle. Surprising finds like these are great for the public and inspiring for people in the field of architecture/urban design.
>
> They actually helped me to stop, pause, and reflect mindfully while on my way to an appointment. Thank you.
>
> Really can lift the energy of a dead space.
>
> Should be more of it! Love them.
>
> We need more projects like this throughout Fremantle. I stumbled across this project today as I walked through Fremantle and was very excited to see something like this popping up in an unexpected place.

Less positive comments tended to focus on the small size of the walls and their subsequent lack of impact, or lack of maintenance.

> The green walls are a great addition to the areas but they are relatively small.
>
> A small step in the right direction. Size reduces the potential impact. The glare from Westgate Mall wall, makes it unpleasant to look for long, and the Mall is very underused due to the risks from vagrant community. Cantonment is a bit lost in the shade of the building. I'd like to see WA native plantings and how they survive ...
>
> It's a great concept but I'm not sure that these walls are big enough to achieve their potential benefits to both people and the environment.
>
> This trial is too small in area. I reckon it needs to be higher and taller. Good to see the species have been changed to better survive in WA's harsh climate. It was looking quite shabby. If this project is the experiment, where is a good sized green wall going to appear? Keep up the good work.
>
> The negativity of my [survey responses] is reflected in the lack of maintenance of these otherwise wonderful projects. Someone desperately needs to either apply water to the one behind Target, or something to protect the plants from Western Australia's harsh sunlight.

Almost all respondents agreed that they wanted to see more biodiversity in cities and that green walls helped preserve nature (Table 5.5). Several people commented on the importance of nature in the city and future sustainability:

Table 5.5 Nature in the city

I would like to see greater biodiversity in cities	**78%**	**21%**	**0.0%**	**1%**	**0.0%**	
	59	*16*	*0*	*1*	*0*	*76*
Green walls can help preserve nature	**54%**	**38%**	**1%**	**2%**	**5%**	
	40	*28*	*1*	2	*4*	*75*

> Contact with nature is important and it's getting increasingly harder to achieve with our population more urbanised.
>
> I am concerned for the future of suburbs and cities as more and more urban bushland is cleared for developments… I am all for preservation of natural environments and for time spent in outdoor natural environments.
>
> Incorporating green walls in Fremantle is another way Fremantle is being recognized as being focused on sustainable outcomes. I think it makes the area a more appealing and welcoming place to be. Hopefully this project will set an example for the rest of Perth, so in the future the city as a whole can be a greener more sustainable city.

Even so, a small proportion of respondents disagreed (4%) or did not know (5%) whether green walls helped to preserve nature. Some comments also questioned the sustainability of the green walls:

> Still feels like a nature Band-Aid.
>
> I dunno if filling a wall with potting mix from a factory is really the most productive way to live sustainably.

It was obvious from the responses that most people enjoy the beauty of nature (Table 5.6). Only one respondent disagreed. It was also apparent that most respondents agreed that being out in nature assisted to reduce stress. Similarly, most respondents agreed that seeing and spending time in nature was important to how they felt during the day and their state of happiness. One respondent commented:

> Nature can provide cognitive benefits. I find it helpful to restore focus.

One respondent commented on their personal experience:

> I agree with having these surprising elements of discovery and places to visit throughout your day and year!! They mark time. I used to visit certain plants in spare blocks throughout the year as a kid … You make friends with plants!!

Most people responded positively to spending time in nature, either in their gardens or enjoying outdoor activities, or seeking out nature experiences in the city (Table 5.7). One comment about spending time in nature as a child highlighted the potential benefits received.

> I grew up when and where bush was all around the suburbs… and swamps and back lanes. It's hard to describe what that kind of freedom gives one.. It's a very different internal works that sustains you throughout your life and is why I think of the importance of these projects for future health if human mind body and spirit.

Table 5.6 Connection to nature

	Strongly agree	Agree	Disagree	Strongly disagree	Don't know	
I enjoy the beauty of nature	**83%**	**16%**	**0%**	**1%**	**0%**	
	62	*12*	*0*	*1*	*0*	*75*
Being out in nature is a great stress reducer for me	**79%**	**20%**	**1%**	**0%**	**0%**	
	58	*15*	*1*	*0*	*0*	*74*
It makes [a] no difference whether I see nature on a daily basis or not	**64%**	**24%**	**5%**	**7%**	**0%**	
	48	*18*	*4*	*5*	*0*	*75*
I need time in nature to be happy	**55%**	**36%**	**5%**	**1%**	**3%**	
	41	*27*	*4*	*1*	*2*	*75*

Table 5.7 Time spent in nature

	Strongly agree	Agree	Disagree	Strongly disagree	Don't know	
I enjoy spending time in my garden	**53%**	**40%**	**6%**	**1%**	**0.0%**	
	39	*29*	*5*	*1*	*0*	*74*
I enjoy spending time in nature (going bushwalking or camping) just for the sake of it	**51%**	**43%**	**5%**	**1%**	**0.0%**	
	38	*32*	*4*	*1*	*0*	*75*
I would plan my walking or cycling route to use streets with green walls	**42%**	**45%**	**4%**	**3%**	**6%**	
	32	*34*	*3*	*2*	*5*	*76*
I seek out nature in cities wherever I can	**41%**	**43%**	**12%**	**3%**	**1%**	
	30	*32*	*9*	*2*	*1*	*74*
I visit local parks as often as I can	**33%**	**44%**	**16%**	**3%**	**4%**	
	25	*33*	*12*	*2*	*3*	*75*

However, this question theme generated the greatest proportion of disagreement. Not all respondents agreed that they enjoyed spending time in their garden (7% disagreed) or in nature (6% disagreed). Similarly, a small proportion disagreed they would actively spend time in nature by planning walking or cycling routes with green walls (7% disagreed and 6% did not know), seeking out nature in the city (15% disagreed and 1% did not know) or visit local parks (19% disagreed and 4% did not know).

Nature connection and response to green walls
How responses to questions about nature influenced overall responses needs to be considered. If respondents were less connected to nature, or did not enjoy spending time outdoors, would they respond as positively to the green walls? Were they more likely to be indifferent?

Recreating green walls
From several comments, it was clear that the walls inspired people to think about creating something similar in their own homes.

> An inspiring initiative. I would like to know more about them, in terms of trialling in my own garden and property, in order for more exposure for the concept in a domestic setting.
>
> I think the green walls are amazing – both the one outside the Fremantle library and the ones on the walls opposite the Woolstores shopping centre. I was inspired by these living visions of green and would like to recreate these at my own house if that is possible. I would love to have the know-how to recreate these at home. More please!! When I first saw them I was stopped in my tracks to look at them and see how they work, what are they made from, how are they watered, etc. They are great!
>
> The plants are thriving and look so healthy and this is April 2015. Would you please provide on your website a list of the plants used in the Fremantle green walls. As I live in the Fremantle area I feel inspired. Could we have some more green walls in Fremantle please? There are many bleak areas in Freo full of hard surfaces without a bit of nature in sight.
>
> Perhaps [City of Fremantle] could make some of their money back by running affordable classes in how to construct green walls for people's homes?
>
> Provide the instructions please put up on the Freo city website.

The one word
Respondents were asked for one word to describe their feelings when they viewed the green wall. These were then aggregated into a word cloud as in Fig. 5.8. 'Inspiring', as the most frequently used word, was possibly influenced by the use of the word in the survey. Many people thought the wall too small, maybe not understanding that it was just a trial. The words were positive, reflecting the overall positive public response to the survey.

Survey conclusions
There was a strong positive response to the function and design benefits which green walls can contribute to a city. This reflected the perception that green walls can enhance the liveability of a city through both their attractiveness and social and environmental amenity. Almost all responses wanted greater biodiversity in cities and thought that green walls could contribute towards this and preserving nature.

Fig. 5.8 The one word (*Source* Author)

Fig. 5.9 The north wall Site A towards the conclusion of the trial (*Source* Author)

Most respondents appreciated the beauty of the walls and the ability of the green wall to make an area more attractive. The enjoyment of the beauty of nature was strongly agreed to, receiving a slightly higher positive response than the recognition that time in nature reduced stress or increasing happiness. These factors, though, were all considered important. To most people, daily contact with nature was important to their sense of well-being. The green walls predominantly elicited a positive emotional response with feelings of wonder and inspiration.

Although the survey sample size and statistical reliability were not sufficient for inferential statistical analysis, it did provide enough for a qualitative descriptive analysis. In this context, there is a reflection of awareness of both social and environmental benefits of green walls and nature (Fig. 5.9).

5.5 Conclusion

The trial went longer than first anticipated.

It was initially agreed between the key players and the building's owners that the walls would be installed for a year. Six years later they are still in place and thriving, bringing beauty and life to otherwise barren areas. Possibly this is why they haven't been taken down as initially planned. In this regard, they are a success. Socially, as shown in the survey, they are appreciated and receive good response. Vandalism hasn't been a problem. Insects are attracted to them and the spring flowers attract the bees (Fig. 5.10).

Fig. 5.10 Springtime and dragonflies (*Source* Author)

References

Battellino R (2010) Mining booms and the Australian economy. RBA Bull (March) 63–69

Bryman A, (2008) Social research methods. Oxford University Press, Oxford, UK

Curtis C (2008) Evolution of the transit-oriented development model for low density cities: a case study of Perth's new railway corridor. Plan Pract Res 23(3):285–302

Daniel G, (2012) Curtin helps the walls of Fremantle come alive. Retrieved from http://news.curtin.edu.au/media-releases/curtin-helps-the-walls-of-fremantle-come-alive/

Moran A, Novak J (2009) The great lock out: the impact of housing and land regulations in Western Australia. Institute of Public Affairs Occasional Paper

Roots and Shoots 11th May 2013, Lush J, Hahn S (Directors) (2013) Roots and Shoots: 720 ABC Perth

Water Corporation (2009) Water forever: towards climate resilience. 2015. https://www.watercorporation.com.au/~/media/files/about-us/planning-for-the-future/water-forever-50-year-plan.pdf

Worthington AC (2012) The quarter century record on housing affordability, affordability drivers and government policy responses in Australia. Int J Hous Mark Anal 5(3):235–252

Ye Q (2008) Commodity booms and their impacts on the Western Australian economy: the iron ore case. Resour Policy 33(2):83–101

Chapter 6
The Progression of Biophilic Design

Abstract Biophilic design is an evolution of human creativity and thinking. Since the early use of the term biophilia in the 1960s, it took some decades to emerge as the design approach known as biophilic design. Birthed by passionate forward thinkers with the resulting seminal book Biophilic Design in 2008, biophilic design has rapidly expanded on a global scale. It is driven by need, passion and recognition of the multiple benefits that ensue. For many, it is now an established, and vital, urban design criterion. The growing Biophilic Cities network has linked global cities and allowed exchanges of knowledge, research and support.

Keywords Biophilic cities · Biophilic design · Biophilic urbanism · Social movement · New York High Line

6.1 Introduction

This concluding chapter brings the biophilic design social movement story up to date. It briefly examines the progression of the movement of biophilic design in recent times and where there are indicators it is moving into the mainstream. It concludes with a personal reflection on the learning throughout the journey and the writing of the thesis.

6.2 The Progression of the Social Movement

It is 35 years since Wilson's book on biophilia, 16 years since *The Biophilia Hypothesis* was published and 11 years since the publication of *Biophilic Design*. The Rhode Island conference in 2006 brought together a diverse but dynamic group whose creative discussion and insights resulted in the birth of biophilic design. Since the book *Biophilic Design* was published, the social movement of biophilic design, as a new approach to the human–nature connection and city design, has rapidly expanded around the world. It has emerged through local champions, collaboration and education.

J. Söderlund, *The Emergence of Biophilic Design*, Cities and Nature,
https://doi.org/10.1007/978-3-030-29813-5_6

Green roofs, as the research literature suggested, are the most implemented biophilic design principle. They are an accepted and expected design feature in Germany and Switzerland. They are mainstreamed in Toronto; where a local innovator of green roofs, noted that from being asked frequently to present informative talks on green roofs, he was no longer invited as the need had passed. Green roofs had been accepted within both government and the community.

The mainstreaming of biophilic design is more than acceptance of green roofs; it is a transition of thinking, an integration of disciplines, where the philosophy of biophilic design is appearing in the everyday approach to urban design and planning. Globally the social movement is at different stages; it greatly varies between cities and between countries. There is the opportunity to learn from the leading cities, to examine their motivators, to tap into presented potential and to fast-track initiatives. A landscape architecture course in Toronto includes biophilic design, as do some university courses in Australia. I personally lecture on biophilic design, with approaches from urban planning and architecture students for information.

In the last 11 years since *Biophilic Design* was published, there has been significant progression in the social movement of biophilic design, driven by a variety of motivators including a connection to nature. Throughout the research discussed in this book, behind the continuous positive comments and interest, there appeared to be a deeper hunger that has been awakened in urbanites: a desire for nature and the forms and patterns of nature to become an integral part of city design available to all.

6.2.1 Biophilic Cities Network

At the time of this book's research, nine cities were signed up as 'biophilic partner cities' with the global biophilic cities network. Since then, many other cities have joined, including Fremantle, Western Australia. As seen in the following graphic (Fig. 6.1) from the Biophilic Cities website, there are now 17 partner cities, with early 2019 seeing more European cities express interest in becoming a partner city. The Biophilic Cities website (biophiliccities.org) has developed, changed and expanded. There have been short documentary films made about biophilic initiatives. Webinars

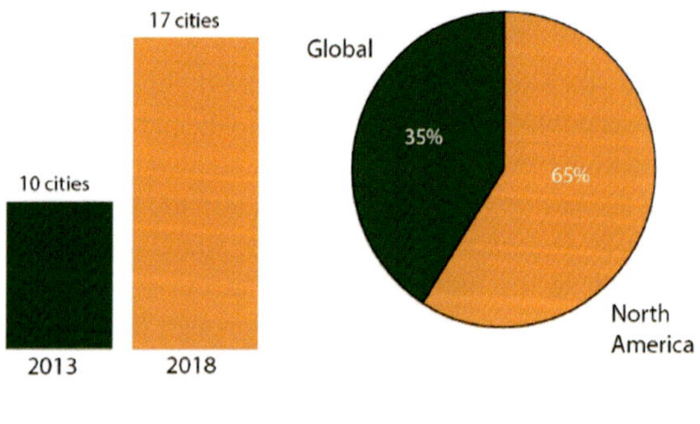

Fig. 6.1 Biophilic partner cities progression (*Source* biophilicities.org)

have been held. Biophilic law and policy has been investigated, and a Biophilic Cities Index has been developed to provide a metric for assessing partner cities progress. Partner cities are also working together to compile a library of metrics to measure abundance and accessibility of nature.

In an interview in Portland, the criteria for quality of life were discussed, and the interviewee concluded that they all fell under the same umbrella. He saw the Biophilic Cities Network as a way of uniting and integrating these qualities. There is a huge potential to do this. The collaboration between city representatives brings together creative minds and clever thinking. Biophilic City conferences could be held globally with greater frequency as the movement expands.

Since this 2013 conversation, this has come to fruition. As an organisation, Biophilic Cities has fostered global collaborations. Biophilic Cities kindly provided a brief update outlining their achievements and direction:

> The breadth of the Biophilic Cities Network has expanded significantly since its inception in 2013. Today it includes a growing and evolving diversity of partner cities, global organizations and individuals. Through the network, participants are communicating about the projects underway in cities and exploring the science and policies that support the planning and design of cities that are rich in accessible nature. Biophilic Cities continues to develop the scholarship that illustrates the benefits of natureful cities through the publication of the Biophilic Cities Journal and a growing library of films and webinars. The unique focus of the network complements other areas of related, critical inquiry such as health, climate adaptation, and equity. The network is connecting the cohort of biophilic cities with major international initiatives to share information and tools with partners like ICLEI, IUCN and The Nature Conservancy. Biophilic Cities is also exploring new opportunities to work collaboratively with global organizations to identify areas of overlap and improved efficiency. Participation in these collective efforts provides Biophilic Cities with broad new partnerships that leverage the connectivity of the Biophilic Cities Network.
>
> (Courtesy of the Biophilic Cities organisation)

In the 6 years since the inaugural conference Biophilic Cities conference, the organisation has grown so significantly that a consultant has been employed to develop a strategic plan. This growth illustrates the importance that people place on the concept and maybe reflects the resonance they feel with it.

There is now a greater understanding of what constitutes a biophilic city.

Biophilic Cities Australia

In August 2017, Tim Beatley again visited Fremantle, Western Australia. During the visit, he presented a talk, followed by a panel discussion. It was extremely well attended and well received, birthing a steering committee whose intention was creating a Biophilic Cities Perth chapter to influence direction and development in Perth. The inaugural meeting of the group happened shortly after. *Biophilic Cities Perth* has held a few events. Film screenings of '*Biophilic Design: The architecture of life*' have been sold out and thus repeated. The group has been invited to participate in events and to present on biophilic cities. A local government organisation requested a workshop for local government public servants which was sold out.

Our *Biophilic Cities Perth* Facebook page has nearly 300 followers and our mailing list has over 500 subscribers.

Fig. 6.2 Biophilic Cities Perth presenting certificate to Mayor of Fremantle and councillors (*Source* Cassie Gunthorpe)

Biophilic Cities Perth has now incorporated, with a name change to *Biophilic Cities Australia*. It was felt that this was a good direction as potentially it would unite other chapters that are forming within Australia and facilitate information sharing.

The group assisted Fremantle with signing up to the *Biophilic Cities* Network and organised an event at Fremantle council to present the Mayor with a *Biophilic Cities* partner certificate (Fig. 6.2).

6.2.2 Literature

In 2015 at the time of this research, two recent books on sustainability included chapters on biophilic design. This indicated that biophilic design was potentially being included as a pathway towards sustainability.

One of the books, *Constructing Green: The Social Structures of Sustainability,* was published in 2013 (Henn and Hoffman 2013). It contains two chapters on biophilic design: 'Incorporating biophilic design through living walls: The decision-making process' (Bartczak et al. 2013) and 'Constructing the biophilic community' (Browning 2013). Another book is a beautifully written and presented comprehensive overview of emerging design philosophies called *Designing for Hope: Pathways towards Regenerative Sustainability* (Hes and Du Plessis 2015). It examines the why and how of regenerative design and sustainability. A sizable chapter is dedicated to establishing the need to reconnect with nature, using *The Biophilia Hypothesis* as rationale and biophilic design elements as guiding principles.

Since this time, biophilic design has become well and truly established and linked with sustainable, resilient and regenerative. As a term it is embedded in the literature: books, journals, articles. Biophilic design frequently appears in grey literature in discussion about human health and well-being, often connected to de-stressing and improving mental health. The need for human–nature connection in built environments both indoor and out has become a global conversation.

Academic articles on biophilic design and nature in cities has increased. Researching my university's catalogue for what comes up under the term biophilic design, I found 1,900 results: 873 articles, 510 dissertations, 372 newspaper articles, 65 reviews and 18 books dwindling down to conference papers and my own one thesis. The library statistics indicate the rapid growth of literature. In the last 3 months, there has been 91 new sources acquired, 38 in the last month.

My paper *on Improving mental health in prisons through biophilic design* was top of the library's list, and regular publisher reports indicate it is well read, illuminating the understanding of the connection between exposure to nature and psychological human health (Söderlund and Newman 2017). There are academic papers on biophilic design in children's libraries, corporate interior design, hotel lobbies and spaces for older people, for example. This illustrates the wide-ranging benefits and potential of biophilic design.

In 2016, Beatley published a new book called 'The Handbook of Biophilic City Planning and Design' to which I contributed a section on the trial green walls in Fremantle (Söderlund in Beatley 2016).

There is a biannual Biophilic Cities Journal which includes spotlights on city projects, academic research, blogs and book reviews.

As implementation of biophilic design initiatives, case studies and research continue to build, so will the literature until, possibly, it is such an integrated component of design, that the need to write about it dissipates.

6.2.3 Conferences

As discovered in the interviews, conferences are an important component of progression of the social movement of biophilic design. I receive regular notification of conferences concerning biophilic design or the principles of biophilic design via email. These are occurring around the globe and continue to unite interest groups in collaboration and education. The CitiesAlive conference still occurs annually. The International Green Roof Congress hold annual conferences. Singapore hosts an annual International Skyrise Greenery Conference. I was invited to keynote a China-ASEAN symposium in Singapore, presenting on biophilic design. Again in Singapore, the International Federation of Landscape Architects held their World Congress in 2018 with a biophilic design theme. I was invited to present my paper on biophilic design in prisons, and from there have been invited to present on biophilic design at a Biophilic Summit held by the Public Health department in Bay of Plenty in New Zealand. This is in conjunction with presentations locally and nationally. Beatley is consistently being invited to keynote at many global conferences.

The third Biophilic Leadership summit is about to be held at Serenbe in the United States. Biophilic Cities is presenting at the Living Futures Institute's unconference. In May 2019, there will be a 'The Nature of Cities Summit' in Paris in June. Biophilic Cities will host a workshop for cities interested in joining the network.

The second Biophilic Cities Network summit is planned to be held in Singapore in October 2019. The Biophilic Cities project which was launched at the conference in 2013 by Tim Beatley continues to grow.

6.2.4 Building Standard Rating Tools

International WELL Building Institute (IWBI) is a corporation whose mission is to improve human health and well-being through the built environment. They introduced the WELL building certification which has two biophilic components, one quantitative and one qualitative. Globally, there are currently 159 WELL certified projects with 1420 WELL registered projects.

In Australia, there is a Green Star rating tool for the built environment. I am advising on its current revision to include more of nature and biophilic design credits.

There is also an increasing number of tools available which assess greenery in urban environments or are used to calculate replacement greenery for a development. These include Singapore's Green Plot Ratio, Seattle's Green Factor and recently London's Urban Greening Factor.

6.2.5 The Influence of the New York High Line

The New York High Line is an example of the success of a biophilic initiative and of the ripple potential of the design approach. With its third stage being built and attracting nearly five million visitors in a year (Grynbaum 2014), the High Line has inspired similar constructions around the world. Sydney, Australia, has begun work on the Goods Line which also utilises a disused rail freight corridor to create a pedestrian green corridor (Chua 2014). Similarly, Chicago's Bloomingdale Trail is attracting funds and attention from government and community to upgrade the trial as has happened with the New York High Line (Sherman 2008). London, Mexico City and Seoul all have similar plans for their own High Lines (Cilento 2011; Kim 2014; Kwaak 2014).

Washington DC is underway on a project called the 11th Street Bridge Park project. This is a linear park spanning the bridge across the Anacostia River which connects two different socio-economic communities. Already riverside improvements have facilitated unity and lessened the divide, yet this project, by including significant community consultation, hopes to improve it even more.

6.2.6 *Social Media and the Professions*

Configurations of biophilic and biophilia have found their way into social media as Twitter and Instagram account names. Being active in Twitter, I find a steady stream of biophilic design-related articles posted and discussed. There is great information sharing. LinkedIn is similar with groups dedicated to nature in cities or biophilic design such as 'Biophilic Design: realizing our need for nature at work (#biophilicoffice)', 'The Biophilic Cities Project' and 'Sustainable, "Green" Design Group'. Searching LinkedIn reveals biophilic design companies, biophilic institutes, biophilic consultants and biophilic architects.

There are bloggers and vloggers who write and film about biophilic design and the human–nature connection. There are Facebook pages such as our own Biophilic Cities Perth.

It appears that the terms biophilic or biophilic design has given many people a name with which they identify, a name to describe their practice and understanding, or inner knowing, to foster a human–nature connection wherever they live.

6.2.7 *Google Trends*

The increasing Internet traffic, although not in the academic arena, is significant as indicating the increasing popular interest in biophilic design principles globally. Google has a site which graphs the search traffic for particular terms.

At the time of the initial research, Google Trends had revealed interest in biophilic design beginning late in 2012, with peaks and troughs from there on (see Appendix B). Improvements from Google in data analysis now shows the first search beginning in 2006 at the time of the Rhode Island conference with a later surge when the *Biophilic Design* book was published. It has been since late 2013 that a steadily increasing search interest has occurred (Fig. 6.3).

6.3 Conclusions

There is no standard way of designing biophilically; the rich palette of design elements and multiple benefits extends the concept across multiple disciplines beyond landscape architects to include biologists, psychologists, architects, urban designers, health professionals into all business and all levels of government. Biophilic design is a significant social movement that is already impacting on urban design and provides

Fig. 6.3 Biophilic design in Google Trends (*Source* Google Trends)

further potential to provide the opportunity and choice for urban dwellers to connect with nature in healthier more liveable cities. While the term biophilia originated through exploration of human nature (psychology), biophilic design is evolving to represent a weaving together of the dichotomy that has existed between humans and nature in the built environment.

The necessary components for biophilic design to integrate into good city design are assembled in books, such as this, research papers and articles. Biophilic design initiatives are an effective response to urban crises, both social and environmental, so there is a need. The technology has been developed to implement, design and to monitor the social and environmental benefits which can then be translated into quantifiable economic benefits. Yet the wholistic picture of these possibilities has not been realised. The gap between community and government, stakeholders and research is contributing to retardation of implementation as both knowledge and practice are fractured. Integration may be assisted through unification of the components under the umbrella term of biophilic design.

Biophilic design is providing a meeting place for humans and nature plus function, beauty, technology, creativity, where top-down can meet bottom-up. It is providing a framework for integration of siloed arenas and benefits and where recognition of the interdependency of the motivators can occur. At the core is a love of life and a human–nature connection, one that includes creativity and peace. Does the rapid uptake of biophilic design where it has been introduced suggest true biophilia, the seeking to unite? Perhaps this increased connection with nature nurtures biophilic tendencies and does help bring nature into humanity and humanity into nature. Biophilic design has the opportunity to define a new relationship with nature and potentially one which is driven by biophilia, not just by the benefits nature brings.

People want nature, the ripple effect demonstrates this. Biophilic design can contribute towards creating cities where people have the opportunity to de-stress, be healthier and more creative. A local champion cannot change the built environment in isolation. They need to find others with the same vision and passion to change or progress. Biophilic design is a global movement and worldwide there are local champions and small groups that are collectively contributing to this greater global movement. They join together through collective rites such as conferences. They build the research. They implement the ideas and concepts which then aids further research. They make it happen. Government policy needs to enable and support the process of positive self-organising communities and provide opportunities for education and collaboration. Integrating the components provides the opportunity for synergistic momentum in biophilic design initiatives, resulting in healthy, liveable and resilient cities where nature, and all its elements, is an integral, accepted and expected inhabitant. Buildings can become more than technological feats, they can be habitats. All urbanites should have access to nature in their lives, the chance to connect with nature or, in Fromm's view, the opportunity to follow the progressive path of life through love. Imagine cities where every day people can interact with nature, where opioid receptors, our molecules of emotion, start zinging, biophilic love is stirred, and the world is a little better for it.

References

Bartczak C, Dunbar B, Bohren L (2013) Incorporating biophilic design through living walls: the decision-making process. In: Henn R, Hoffman A (eds) Constructing green: the social structures of sustainability. Massachusetts Institute of Technology, Massachusetts

Browning B (2013) Constructing the biophilic community. In: Henn R, Hoffman A (eds) Constructing green: The social structures of sustainability. Massachusetts Institute of Technology, Massachusetts

Chua G (2014) New design concepts unveiled for Sydney's New York style 'high line'. Architecture and Design. http://www.architectureanddesign.com.au/home

Cilento K (2011) Mexico City's high line park. Arch Daily. http://www.archdaily.com/162420/mexico-citys-high-line-park/

Grynbaum M (2014). Highline draws millions but de Blasio isn't one of them. The New York Times. http://www.nytimes.com/2014/10/02/nyregion/high-line-draws-millions-but-de-blasio-isnt-one.html

Hes D, Du Plessis C (2015) Designing for hope: pathways to regenerative sustainability. Routledge, Oxon, UK

Kim J (2014) Will London's garden bridge rival New York's high line, or is it just a tacky tourist trap? Quartz. http://qz.com/307475/will-londons-garden-bridge-rival-new-yorks-high-line-or-is-it-just-a-tacky-tourist-trap/

Kwaak J (2014) Seoul plans high line style elevated park. http://blogs.wsj.com/korearealtime/2014/09/01/seoul-plans-high-line-style-elevated-park/

Sherman D (2008) Chicago's park in the sky: the Bloomingdale trail. http://www.thehighline.org/blog/2008/02/13/chicagos-park-in-the-sky-the-bloomingdale-trail

Söderlund J (2016) Piloting green walls and studying their effects. In: Beatley (ed) Handbook of biophilic city planning and design. Island Press, Washington, DC, pp 156–157

Söderlund J, Newman P (2017) Improving mental health in prisons through Biophilic Design. Prison J (Sage Publishing). http://journals.sagepub.com/eprint/vBW5GG7kibawPAEqA2sD/full

Appendix A
Analysis Tables

A.1 Analysis of Design Principles

See Table A.1.

A.2 Analysis of Researched Human–Nature Benefits

See Table A.2.

J. Söderlund, *The Emergence of Biophilic Design*, Cities and Nature,
https://doi.org/10.1007/978-3-030-29813-5

Table A.1 Meta-analysis of design principles

Design principle	Benefit	Social	Environmental	Economic	Country of study	References
Vertical living walls						
	Aesthetically pleasing	X[1], O[2]			[1]U.K., [2]Egypt	[1]White, Emma V., and Birgitta Gatersleben. 2011. "Greenery on Residential Buildings: Does It Affect Preferences and Perceptions of Beauty?" *Journal of Environmental Psychology* 31 (1): 89–98. http://dx.doi.org/10.1016/j.jenvp.2010.11.002 [2]Sheweka, Samar Mohamed, and Nourhan Magdy Mohamed. 2012. "Green Facades as a New Sustainable Approach Towards Climate Change." Energy Procedia 18 (0): 507–520. http://dx.doi.org/10.1016/j.egypro.2012.05.062
	Psychologically restorative	X[1]			[1]U.K.	[1]White, Emma V., and Birgitta Gatersleben. 2011. "Greenery on Residential Buildings: Does It Affect Preferences and Perceptions of Beauty?" *Journal of Environmental Psychology* 31 (1): 89–98. http://dx.doi.org/10.1016/j.jenvp.2010.11.002
	Noise attenuation		X[1], O[2]		[1]Singapore, [2]Egypt	[1]Wong, Nyuk Hien, Alex Yong Kwang Tan, Puay Yok Tan, Kelly Chiang, and Ngian Chung Wong. 2010. "Acoustics Evaluation of Vertical Greenery Systems for Building Walls." *Building and Environment* 45 (2): 411–420. http://dx.doi.org/10.1016/j.buildenv.2009.06.017 [2]Sheweka, Samar Mohamed, and Nourhan Magdy Mohamed. 2012. "Green Facades as a New Sustainable Approach Towards Climate Change." Energy Procedia 18 (0): 507–520. http://dx.doi.org/10.1016/j.egypro.2012.05.062
	Life cycle analysis		X[1]		[1]Netherlands	[1]Ottelé, Marc, Katia Perini, A. L. A. Fraaij, E. M. Haas, and R. Raiteri. 2011. "Comparative Life Cycle Analysis for Green Façades and Living Wall Systems." *Energy and Buildings* 43 (12): 3419–3429. http://dx.doi.org/10.1016/j.enbuild.2011.09.010
	Increased building lifespan		O[1]		[1]Egypt	[1]Sheweka, Samar Mohamed, and Nourhan Magdy Mohamed. 2012. "Green Facades as a New Sustainable Approach Towards Climate Change." Energy Procedia 18 (0): 507–520. http://dx.doi.org/10.1016/j.egypro.2012.05.062
	Water recycling		X[1]		[1]U.S.A.	[1]Burrows, R.M., and M.A. Corragio. 2011. "Living Walls: Integration of Water Re-Use Systems." In *Cities Alive: 9th Annual Green Roof and Wall Conference, Philadelphia,* edited by Green Roofs for Healthy Cities. **Green Wall water re-use irrigation*

(continued)

Table A.1 (continued)

Design principle	Benefit	Social	Environmental	Economic	Country of study	References
	Decreased energy consumption		X[1], X[2], X[3], O[5]	O[1*]	[1]Netherlands, [2]Greece, [3]Singapore, [4]Spain, [5]Egypt	[1]Ottelé, Marc, Katia Perini, A. L. A. Fraaij, E. M. Haas, and R. Raiteri. 2011. "Comparative Life Cycle Analysis for Green Façades and Living Wall Systems." *Energy and Buildings* 43 (12): 3419–3429. http://dx.doi.org/10.1016/j.enbuild.2011.09.010 [2]Kontoleon, K. J., and E. A. Eumorfopoulou. 2010. "The Effect of the Orientation and Proportion of a Plant-Covered Wall Layer on the Thermal Performance of a Building Zone." *Building and Environment* 45 (5): 1287–1303. http://dx.doi.org/10.1016/j.buildenv.2009.11.013 [3]Wong, Nyuk Hien, Alex Yong Kwang Tan, Puay Yok Tan, and Ngian Chung Wong. 2009. "Energy Simulation of Vertical Greenery Systems." *Energy and Buildings* 41 (12): 1401–1408. http://dx.doi.org/10.1016/j.enbuild.2009.08.010 [4]Pérez, G., L. Rincón, A. Vila, J. M. González, and L. F. Cabeza. 2011. "Behaviour of Green Facades in Mediterranean Continental Climate." *Energy Conversion and Management* 52 (4): 1861–1867. http://dx.doi.org/10.1016/j.enconman.2010.11.008. *(inconclusive) [5]Sheweka, Samar Mohamed, and Nourhan Magdy Mohamed. 2012. "Green Facades as a New Sustainable Approach Towards Climate Change." Energy Procedia 18 (0): 507–520. http://dx.doi.org/10.1016/j.egypro.2012.05.062
	Passive interior cooling	O[3]	X[1], X[2], X[3], X[4], X[5], X[6]	O[3]	[1]Greece, [2]Singapore, [3]Egypt, [4]Hong Kong, [5]Egypt, [6]Spain	[1]Kontoleon, K. J., and E. A. Eumorfopoulou. 2010. "The Effect of the Orientation and Proportion of a Plant-Covered Wall Layer on the Thermal Performance of a Building Zone." *Building and Environment* 45 (5): 1287–1303. http://dx.doi.org/10.1016/j.buildenv.2009.11.013 [2]Wong, Nyuk Hien, Alex Yong Kwang Tan, Puay Yok Tan, and Ngian Chung Wong. 2009. "Energy Simulation of Vertical Greenery Systems." *Energy and Buildings* 41 (12): 1401–1408. http://dx.doi.org/10.1016/j.enbuild.2009.08.010 [3]Sheweka, Samar, and Arch Nourhan Magdy. 2011. "The Living Walls as an Approach for

(continued)

Table A.1 (continued)

Design principle	Benefit	Social	Environmental	Economic	Country of study	References
						a Healthy Urban Environment." *Energy Procedia* 6 (0): 592–599. http://dx.doi.org/10.1016/j.egypro.2011.05.068 [4]Cheng, C. Y., Ken K. S. Cheung, and L. M. Chu. 2010. "Thermal Performance of a Vegetated Cladding System on Facade Walls." *Building and Environment* 45 (8): 1779–1787. http://dx.doi.org/10.1016/j.buildenv.2010.02.005 [5]Sheweka, Samar Mohamed, and Nourhan Magdy Mohamed. 2012. "Green Facades as a New Sustainable Approach Towards Climate Change." Energy Procedia 18 (0): 507–520. http://dx.doi.org/10.1016/j.egypro.2012.05.062 [6]Pérez, G., L. Rincón, A. Vila, J. M. González, and L. F. Cabeza. 2011. "Behaviour of Green Facades in Mediterranean Continental Climate." *Energy Conversion and Management* 52 (4): 1861–1867. http://dx.doi.org/10.1016/j.enconman.2010.11.008. *(inconclusive)
	Improved air quality		X[1], X[2], O[3]		[1]U.K., [2]Egypt, [3]Egypt	[1]Pugh, Thomas A. M., A. Robert MacKenzie, J. Duncan Whyatt, and C. Nicholas Hewitt. 2012. "Effectiveness of Green Infrastructure for Improvement of Air Quality" [2]Sheweka, Samar, and Arch Nourhan Magdy. 2011. "The Living Walls as an Approach for a Healthy Urban Environment." *Energy Procedia* 6 (0): 592–599. http://dx.doi.org/10.1016/j.egypro.2011.05.068 [3]Sheweka, Samar Mohamed, and Nourhan Magdy Mohamed. 2012. "Green Facades as a New Sustainable Approach Towards Climate Change." Energy Procedia 18 (0): 507–520. http://dx.doi.org/10.1016/j.egypro.2012.05.062
	Urban heat island effect mitigation		X[1], X[2], X[3]	O[2]	[1]Singapore, [2]Egypt, [3]Egypt	[1]Wong, Nyuk Hien, Alex Yong Kwang Tan, Puay Yok Tan, and Ngian Chung Wong. 2009. "Energy Simulation of Vertical Greenery Systems." *Energy and Buildings* 41 (12): 1401–1408. http://dx.doi.org/10.1016/j.enbuild.2009.08.010 [2]Sheweka, Samar, and Arch Nourhan Magdy. 2011. "The Living Walls as an Approach for a Healthy Urban Environment." *Energy Procedia* 6 (0): 592–599. http://dx.doi.org/10.1016/j.egypro.2011.05.068

(continued)

Table A.1 (continued)

Design principle	Benefit	Social	Environmental	Economic	Country of study	References
						[3]Sheweka, Samar Mohamed, and Nourhan Magdy Mohamed. 2012. "Green Facades as a New Sustainable Approach Towards Climate Change." Energy Procedia 18 (0): 507–520. http://dx.doi.org/10.1016/j.egypro.2012.05.062
	Wind protection		X[1]		[1]Egypt	[1]Sheweka, Samar Mohamed, and Nourhan Magdy Mohamed. 2012. "Green Facades as a New Sustainable Approach Towards Climate Change." Energy Procedia 18 (0): 507–520. http://dx.doi.org/10.1016/j.egypro.2012.05.062
	Increased building rating		O[1]		[1]Egypt	[1]Sheweka, Samar Mohamed, and Nourhan Magdy Mohamed. 2012. "Green Facades as a New Sustainable Approach Towards Climate Change." Energy Procedia 18 (0): 507–520. http://dx.doi.org/10.1016/j.egypro.2012.05.062
	Building value			O[1]	[1]Egypt	[1]Sheweka, Samar Mohamed, and Nourhan Magdy Mohamed. 2012. "Green Facades as a New Sustainable Approach Towards Climate Change." Energy Procedia 18 (0): 507–520. http://dx.doi.org/10.1016/j.egypro.2012.05.062
Green roofs						
	Aesthetically pleasing	X[1], X[2]			[1]U.K., [2]U.S.A	[1]White, Emma V., and Birgitta Gatersleben. 2011. "Greenery on Residential Buildings: Does It Affect Preferences and Perceptions of Beauty?" *Journal of Environmental Psychology* 31 (1): 89–98. http://dx.doi.org/10.1016/j.jenvp.2010.11.002 [2]Jungels, Jeremy, Donald A. Rakow, Shorna Broussard Allred, and Sonja M. Skelly. 2013. "Attitudes and Aesthetic Reactions toward Green Roofs in the Northeastern United States." *Landscape and Urban Planning* 117 (0): 13–21. http://dx.doi.org/10.1016/j.landurbplan.2013.04.013
	Psychologically restorative	X[1]			[1]U.K.	[1]White, Emma V., and Birgitta Gatersleben. 2011. "Greenery on Residential Buildings: Does It Affect Preferences and Perceptions of Beauty?" *Journal of Environmental Psychology* 31 (1): 89–98. http://dx.doi.org/10.1016/j.jenvp.2010.11.002

(continued)

Table A.1 (continued)

Design principle	Benefit	Social	Environmental	Economic	Country of study	References
	Storm water management		X[1], X[2], X[3], X[4], X[5], X[6], X[7], X[8], X[9]	O[7]	[1]U.S.A., [2]U.S.A., [3]Belgium, Germany, [4]Sweden, [5]U.S.A., [6]U.K., [7]U.S.A., [8]U.S.A., [9]Belgium, [10]U.S.A., Hong Kong, Malaysia	[1]Schroll, Erin, John Lambrinos, Tim Righetti, and David Sandrock. 2011. "The Role of Vegetation in Regulating Stormwater Runoff from Green Roofs in a Winter Rainfall Climate." *Ecological Engineering* 37 (4): 595–600. http://dx.doi.org/10.1016/j.ecoleng.2010.12.020 [2]Gregoire, Bruce G., and John C. Clausen. 2011. "Effect of a Modular Extensive Green Roof on Stormwater Runoff and Water Quality." *Ecological Engineering* 37 (6): 963–969. http://dx.doi.org/10.1016/j.ecoleng.2011.02.004 [3]Mentens, Jeroen, Dirk Raes, and Martin Hermy. 2006. "Green Roofs as a Tool for Solving the Rainwater Runoff Problem in the Urbanized 21st Century?" *Landscape and Urban Planning* 77 (3): 217–226. http://dx.doi.org/10.1016/j.landurbplan.2005.02.010 [4]Czemiel Berndtsson, Justyna. 2010. "Green Roof Performance Towards Management of Runoff Water Quantity and Quality: A Review." *Ecological Engineering* 36 (4): 351–360. http://dx.doi.org/10.1016/j.ecoleng.2009.12.014 [5]Rowe, D. Bradley. 2011. "Green Roofs as a Means of Pollution Abatement." *Environmental Pollution* 159 (8–9): 2100–10. https://doi.org/10.1016/j.envpol.2010.10.029 [6]Nagase, Ayako, and Nigel Dunnett. 2012. "Amount of Water Runoff from Different Vegetation Types on Extensive Green Roofs: Effects of Plant Species, Diversity and Plant Structure." *Landscape and Urban Planning* 104 (3–4): 356–363. http://dx.doi.org/10.1016/j.landurbplan.2011.11.001 [7]Wang, Ranran, Matthew Eckelman, and Julie Zimmerman. 2013. "Consequential Environmental and Economic Life Cycle Assessment of Green and Gray Stormwater Infrastructures for Combined Sewer Systems." *Environmental Science & Technology* 47 (19): 11189–11189 [8]Volder, Astrid, and Bruce Dvorak. 2014. "Event Size, Substrate Water Content and

(continued)

Table A.1 (continued)

Design principle	Benefit	Social	Environmental	Economic	Country of study	References
						Vegetation Affect Storm Water Retention Efficiency of an Un-Irrigated Extensive Green Roof System in Central Texas." *Sustainable Cities and Society* 10 (0): 59–64. http://dx.doi.org/10.1016/j.scs.2013.05.005 [9]Vanuytrecht, Eline, Carmen Van Mechelen, Koenraad Van Meerbeek, Patrick Willems, Martin Hermy, and Dirk Raes. 2014. "Runoff and Vegetation Stress of Green Roofs under Different Climate Change Scenarios." *Landscape and Urban Planning* 122 (0): 68–77. http://dx.doi.org/10.1016/j.landurbplan.2013.11.001 [10]Berardi, Umberto, Amir Hosein Ghaffarian Hoseini, and Ali Ghaffarian Hoseini. 2014. "State-of-the-Art Analysis of the Environmental Benefits of Green Roofs." *Applied Energy* 115 (0): 411–428. http://dx.doi.org/10.1016/j.apenergy.2013.10.047
	Water runoff quality		X[1], X[2], X[3], X[4], X[5], X[6]	O[4],	[1]U.S.A., [2]Sweden, [3]U.S.A., [4]U.S.A., [5]France, [6]U.S.A., Hong Kong, Malaysia	[1]Gregoire, Bruce G., and John C. Clausen. 2011. "Effect of a Modular Extensive Green Roof on Stormwater Runoff and Water Quality." *Ecological Engineering* 37 (6): 963–969. http://dx.doi.org/10.1016/j.ecoleng.2011.02.004 [2]Czemiel Berndtsson, Justyna. 2010. "Green Roof Performance Towards Management of Runoff Water Quantity and Quality: A Review." *Ecological Engineering* 36 (4): 351–360. http://dx.doi.org/10.1016/j.ecoleng.2009.12.014 [3]Rowe, D. Bradley. 2011. "Green Roofs as a Means of Pollution Abatement." *Environmental Pollution* 159 (8–9): 2100–10. https://doi.org/10.1016/j.envpol.2010.10.029 [4]Wang, Ranran, Matthew Eckelman, and Julie Zimmerman. 2013. "Consequential Environmental and Economic Life Cycle Assessment of Green and Gray Stormwater Infrastructures for Combined Sewer Systems." *Environmental Science & Technology* 47 (19): 11189–11189 [5]Seidl, Martin, Marie-Christine Gromaire, Mohamed Saad, and Bernard De Gouvello. 2013. "Effect of Substrate Depth and Rain-Event History on the

(continued)

Table A.1 (continued)

Design principle	Benefit	Social	Environmental	Economic	Country of study	References
						Pollutant Abatement of Green Roofs." *Environmental Pollution* 183 (0): 195–203. http://dx.doi.org/10.1016/j.envpol.2013.05.026 [6]Berardi, Umberto, AmirHosein GhaffarianHoseini, and Ali GhaffarianHoseini. 2014. "State-of-the-Art Analysis of the Environmental Benefits of Green Roofs." *Applied Energy* 115 (0): 411–428. http://dx.doi.org/10.1016/j.apenergy.2013.10.047
	Flood mitigation		X[1]		[1]South Korea	[1]Lee, J. Y., H. J. Moon, T. I. Kim, H. W. Kim, and M. Y. Han. 2013. "Quantitative Analysis on the Urban Flood Mitigation Effect by the Extensive Green Roof System." *Environmental Pollution* 181 (0): 257–261. http://dx.doi.org/10.1016/j.envpol.2013.06.039
	Air quality/air pollution reduction		X[1], X[2], X[3], X[4]		[1]U.S.A., [2]U.K., [3]Hong Kong, [4]U.S.A., Hong Kong, Malaysia	[1]Rowe, D. Bradley. 2011. "Green Roofs as a Means of Pollution Abatement." *Environmental Pollution* 159 (8–9): 2100–10. https://doi.org/10.1016/j.envpol.2010.10.029 [2]Pugh, Thomas A. M., A. Robert MacKenzie, J. Duncan Whyatt, and C. Nicholas Hewitt. 2012. "Effectiveness of Green Infrastructure for Improvement of Air Quality" [3]Leung, Dennis Y. C., Jeaniek Y. Tsui, Feng Chen, Wing-kin Yip, Lilianl P. Vrijmoed, and Chun-ho Liu. 2011. "Effects of Urban Vegetation on Urban Air Quality." *Landscape Research* 36 (2): 173–188. https://doi.org/10.1080/01426397.2010.547570 [4]Berardi, Umberto, AmirHosein GhaffarianHoseini, and Ali GhaffarianHoseini. 2014. "State-of-the-Art Analysis of the Environmental Benefits of Green Roofs." *Applied Energy* 115 (0): 411–428. http://dx.doi.org/10.1016/j.apenergy.2013.10.047
	Carbon dioxide reduction		X[1]		[1]U.S.A.	[1]Rowe, D. Bradley. 2011. "Green Roofs as a Means of Pollution Abatement." *Environmental Pollution* 159 (8–9): 2100–10. https://doi.org/10.1016/j.envpol.2010.10.029

(continued)

Table A.1 (continued)

Design principle	Benefit	Social	Environmental	Economic	Country of study	References
	Noise attenuation		X[1], X[2]		[1]U.S.A., [2]U.S.A., Hong Kong, Malaysia	[1]Rowe, D. Bradley. 2011. "Green Roofs as a Means of Pollution Abatement." *Environmental Pollution* 159 (8–9): 2100–10. https://doi.org/10.1016/j.envpol.2010.10.029 [2]Berardi, Umberto, AmirHosein GhaffarianHoseini, and Ali GhaffarianHoseini. 2014. "State-of-the-Art Analysis of the Environmental Benefits of Green Roofs." *Applied Energy* 115 (0): 411–428. http://dx.doi.org/10.1016/j.apenergy.2013.10.047
	Decreased energy consumption		X[1], X[2], X[3], X[4]	O[3]	[1]France, [2]Italy, Spain, Egypt, [3]U.S.A., Hong Kong, Malaysia, [4]U.S.A.	[1]Jaffal, Issa, Salah-Eddine Ouldboukhitine, and Rafik Belarbi. 2012. "A Comprehensive Study of the Impact of Green Roofs on Building Energy Performance." *Renewable Energy* 43 (0): 157–164. http://dx.doi.org/10.1016/j.renene.2011.12.004 [2]Zinzi, M., and S. Agnoli. 2012. "Cool and Green Roofs. An Energy and Comfort Comparison between Passive Cooling and Mitigation Urban Heat Island Techniques for Residential Buildings in the Mediterranean Region." *Energy and Buildings* 55 (0): 66–76. http://dx.doi.org/10.1016/j.enbuild.2011.09.024 [3]Berardi, Umberto, AmirHosein GhaffarianHoseini, and Ali GhaffarianHoseini. 2014. "State-of-the-Art Analysis of the Environmental Benefits of Green Roofs." *Applied Energy* 115 (0): 411–428. http://dx.doi.org/10.1016/j.apenergy.2013.10.047 [4]Susca, T., S. R. Gaffin, and G. R. Dell'Osso. 2011. "Positive Effects of Vegetation: Urban Heat Island and Green Roofs." *Environmental Pollution* 159 (8–9): 2119–2126. http://dx.doi.org/10.1016/j.envpol.2011.03.007
	Passive cooling		X[1], X[2]		[1]Italy, Spain, Egypt, [2]U.S.A.	[1]Zinzi, M., and S. Agnoli. 2012. "Cool and Green Roofs. An Energy and Comfort Comparison between Passive Cooling and Mitigation Urban Heat Island Techniques for Residential Buildings in the Mediterranean Region." *Energy and Buildings* 55 (0): 66–76. http://dx.doi.org/10.1016/j.enbuild.2011.09.024 [2]Susca, T., S. R. Gaffin, and G. R. Dell'Osso. 2011. "Positive

(continued)

Table A.1 (continued)

Design principle	Benefit	Social	Environmental	Economic	Country of study	References
						Effects of Vegetation: Urban Heat Island and Green Roofs." *Environmental Pollution* 159 (8–9): 2119–2126. http://dx.doi.org/10.1016/j.envpol.2011.03.007
	Urban heat island effect		X[1], X[2], X[3], X[4]	O[2]	[1]Italy, Spain, Egypt, [2]U.S.A., Hong Kong, Malaysia, [3]U.S.A., [4]Hong Kong	[1]Zinzi, M., and S. Agnoli. 2012. "Cool and Green Roofs. An Energy and Comfort Comparison between Passive Cooling and Mitigation Urban Heat Island Techniques for Residential Buildings in the Mediterranean Region." *Energy and Buildings* 55 (0): 66–76. http://dx.doi.org/10.1016/j.enbuild.2011.09.024 [2]Berardi, Umberto, AmirHosein GhaffarianHoseini, and Ali GhaffarianHoseini. 2014. "State-of-the-Art Analysis of the Environmental Benefits of Green Roofs." *Applied Energy* 115 (0): 411–428. http://dx.doi.org/10.1016/j.apenergy.2013.10.047 [3]Susca, T., S. R. Gaffin, and G. R. Dell'Osso. 2011. "Positive Effects of Vegetation: Urban Heat Island and Green Roofs." *Environmental Pollution* 159 (8–9): 2119–2126. http://dx.doi.org/10.1016/j.envpol.2011.03.007 [4]Wong, Johnny Kwok Wai, and Leo Siu-Kit Lau. 2013. "From the 'Urban Heat Island' to the 'Green Island'? A Preliminary Investigation into the Potential of Retrofitting Green Roofs in Mongkok District of Hong Kong." *Habitat International* 39 (0): 25–35. http://dx.doi.org/10.1016/j.habitatint.2012.10.005
	Economic savings		O[1]	X[1]	[1]U.S.A.	[1]Sproul, Julian, Man Pun Wan, Benjamin H. Mandel, and Arthur H. Rosenfeld. 2014. "Economic Comparison of White, Green, and Black Flat Roofs in the United States." *Energy and Buildings* 71 (0): 20–27. http://dx.doi.org/10.1016/j.enbuild.2013.11.058
	Biodiversity		X[1], X[2], X[3], X[4]		[1]U.K., [2]U.K., [3]Switzerland, [4]U.S.A., Hong Kong, Malaysia	[1]Grant, Gary. 2006. "Extensive Green Roofs in London." *Urban Habitats* 4, 1 (ISSN 1541–7115) [2]Kadas, G. 2006. "Rare Invertebrates Colonising Green Roofs in London." *Urban Habitats* 4, 1 (ISSN 1541–7115) [3]Brenneisan, S. 2006. "Space for Urban Wildlife: Designing Green Roofs as Habitats in

(continued)

Table A.1 (continued)

Design principle	Benefit	Social	Environmental	Economic	Country of study	References
						Switzerland." *Urban Habitats* 4, Number 1 (ISSN 1541–7115) [4]Berardi, Umberto, AmirHosein GhaffarianHoseini, and Ali GhaffarianHoseini. 2014. "State-of-the-Art Analysis of the Environmental Benefits of Green Roofs." *Applied Energy* 115 (0): 411–428. http://dx.doi.org/10.1016/j.apenergy.2013.10.047
	Life cycle benefit analysis	X[1]		O[1], X[2]	[1]Canada, U.S.A.	[1]Bianchini, Fabricio, and Kasun Hewage. 2012. "Probabilistic Social Cost-Benefit Analysis for Green Roofs: A Lifecycle Approach." *Building and Environment* 58 (0): 152–162. http://dx.doi.org/10.1016/j.buildenv.2012.07.005 [2]Berardi, Umberto, AmirHosein GhaffarianHoseini, and Ali GhaffarianHoseini. 2014. "State-of-the-Art Analysis of the Environmental Benefits of Green Roofs." *Applied Energy* 115 (0): 411–428. http://dx.doi.org/10.1016/j.apenergy.2013.10.047
	Climate and vegetation performance		X[1]		[1]Belgium	[1]Vanuytrecht, Eline, Carmen Van Mechelen, Koenraad Van Meerbeek, Patrick Willems, Martin Hermy, and Dirk Raes. 2014. "Runoff and Vegetation Stress of Green Roofs under Different Climate Change Scenarios." *Landscape and Urban Planning* 122 (0): 68–77. http://dx.doi.org/10.1016/j.landurbplan.2013.11.001
	Policy	X[1]	O[1]	O[1]	[1]U.S.A., Hong Kong, Malaysia	[1]Berardi, Umberto, AmirHosein GhaffarianHoseini, and Ali GhaffarianHoseini. 2014. "State-of-the-Art Analysis of the Environmental Benefits of Green Roofs." *Applied Energy* 115 (0): 411–428. http://dx.doi.org/10.1016/j.apenergy.2013.10.047
Green connectors						
	Increased biodiversity		X[1], X[2]		[1]U.K., [2]France	[1]Angold, P. G., J. P. Sadler, M. O. Hill, A. Pullin, S. Rushton, K. Austin, E. Small et al. 2006. "Biodiversity in Urban Habitat Patches." *Science of The Total Environment* 360 (1–3): 196–204. http://dx.doi.org/10.1016/j.scitotenv.2005.08.035 [2]Savard, Jean-Pierre L., Philippe Clergeau, and Gwenaelle Mennechez. 2000. "Biodiversity Concepts and Urban Ecosystems." *Landscape and Urban Planning* 48 (3–4): 131–142. http://dx.doi.org/10.1016/S0169-2046(00)00037-2

(continued)

Table A.1 (continued)

Design principle	Benefit	Social	Environmental	Economic	Country of study	References
	Increased land value			O[1]	[1]France	[1]Savard, Jean-Pierre L., Philippe Clergeau, and Gwenaelle Mennechez. 2000. "Biodiversity Concepts and Urban Ecosystems." *Landscape and Urban Planning* 48 (3–4): 131–142. http://dx.doi.org/10.1016/S0169-2046(00)00037-2
Bioretention basin						
	Increased water quality (pollution reduction)		X[1]	O[1]	[1]U.S.A.	[1]Wang, Ranran, Matthew Eckelman, and Julie Zimmerman. 2013. "Consequential Environmental and Economic Life Cycle Assessment of Green and Gray Stormwater Infrastructures for Combined Sewer Systems." *Environmental Science & Technology* 47 (19): 11189–11189
	Aquatic environmental benefits		X[1]	O[1]	[1]U.S.A.	[1]Wang, Ranran, Matthew Eckelman, and Julie Zimmerman. 2013. "Consequential Environmental and Economic Life Cycle Assessment of Green and Gray Stormwater Infrastructures for Combined Sewer Systems." *Environmental Science & Technology* 47 (19): 11189–11189
Indoor plants						
	Increased attention capacity	X[1], X[2]			[1]Norway, [2]Netherland, U.K.	[1]Raanaas, Ruth K., Katinka Horgen Evensen, Debra Rich, Gunn Sjøstrøm, and Grete Patil. 2011. "Benefits of Indoor Plants on Attention Capacity in an Office Setting." *Journal of Environmental Psychology* 31 (1): 99–105. http://dx.doi.org/10.1016/j.jenvp.2010.11.005 [2]Nieuwenhuis, Marlon, Craig Knight, Tom Postmes, and S. Alexander Haslam. 2014. "The Relative Benefits of Green Versus Lean Office Space: Three Field Experiments." *Journal of Experimental Psychology: Applied* 20 (3): 199–214. https://doi.org/10.1037/xap0000024
	Stress reduction	X[1], X[2], X[3], O[4]	X[4]		[1]Japan, [2]U.S.A., [3]South Korea, [4]Australia	[1]Ikei, Harumi, Misako Komatsu, Chorong Song, Eri Himoro, and Yoshifumi Miyazaki. 2014. "The Physiological and Psychological Relaxing Effects of Viewing Rose Flowers in Office Workers." *Journal of Physiological Anthropology* 33 (1): 6. http://www.jphysiolanthropol.com/content/33/1/6 [2]Wener, Richard, and Hannah Carmalt. 2006. "Environmental

(continued)

Table A.1 (continued)

Design principle	Benefit	Social	Environmental	Economic	Country of study	References
						Psychology and Sustainability in High-Rise Structures." *Technology in Society* 28 (1–2): 157–167. http://dx.doi.org/10.1016/j.techsoc.2005.10.016 [3]Park, S. H., and R. H. Mattson. 2008. "Effects of Flowering and Foliage Plants in Hospital Rooms on Patients Recovering from Abdominal Surgery." *HortTechnology* 18 (4): 563–568 [4]Daly, John, Margaret Burchett, and Fraser Torpy. 2010. "Plants in the Classroom Can Improve Student Performance." *National Interior Plantscape Association*
	Increased air quality/pollutant removal	O[1], O[3]	X[1], X[2], X[3]		[1]U.S.A., [2]Japan, [3]Portugal	[1]Wener, Richard, and Hannah Carmalt. 2006. "Environmental Psychology and Sustainability in High-Rise Structures." *Technology in Society* 28 (1–2): 157–167. http://dx.doi.org/10.1016/j.techsoc.2005.10.016 [2]Oyabu, Takashi, Ayako Sawada, Takeshi Onodera, Kozaburo Takenaka, and Bill Wolverton. 2003. "Characteristics of Potted Plants for Removing Offensive Odors." *Sensors and Actuators B: Chemical* 89 (1–2): 131–136. http://dx.doi.org/10.1016/S0925-4005(02)00454-9 [3]Pegas, P. N., C. A. Alves, T. Nunes, E. F. Bate-Epey, M. Evtyugina, and C. A. Pio. 2012. "Could Houseplants Improve Indoor Air Quality in Schools?" *Journal of Toxicology and Environmental Health, Part A* 75 (22–23): 1371–1380. https://doi.org/10.1080/15287394.2012.721169
	Increased healing rates	X[1]			[1]South Korea	[1]Park, S. H., and R. H. Mattson. 2008. "Effects of Flowering and Foliage Plants in Hospital Rooms on Patients Recovering from Abdominal Surgery." *HortTechnology* 18 (4): 563–568
	Increased productivity	X[1], X[2]		O[2]	[1]Australia, [2]Netherland, U.K.	[1]Daly, John, Margaret Burchett, and Fraser Torpy. 2010. "Plants in the Classroom Can Improve Student Performance." *National Interior Plantscape Association* [2]Nieuwenhuis, Marlon, Craig Knight, Tom Postmes, and S. Alexander Haslam. 2014. "The Relative Benefits of Green Versus Lean Office Space: Three Field Experiments." *Journal of Experimental Psychology: Applied* 20 (3): 199–214. https://doi.org/10.1037/xap0000024

(continued)

Table A.1 (continued)

Design principle	Benefit	Social	Environmental	Economic	Country of study	References
	Increased positive mood	X[1]			[1]Netherland, U.K.	[1]Nieuwenhuis, Marlon, Craig Knight, Tom Postmes, and S. Alexander Haslam. 2014. "The Relative Benefits of Green Versus Lean Office Space: Three Field Experiments." *Journal of Experimental Psychology: Applied* 20 (3): 199–214. https://doi.org/10.1037/xap0000024
Natural lighting						
	Stress reduction	X[1]			[1]U.S.A.	[1]Wener, Richard, and Hannah Carmalt. 2006. "Environmental Psychology and Sustainability in High-Rise Structures." *Technology in Society* 28 (1–2): 157–167. http://dx.doi.org/10.1016/j.techsoc.2005.10.016
View of nature						
	Attention capacity increase	X[1], X[2]			[1]U.S.A., [2]Netherland, U.K.	[1]Wener, Richard, and Hannah Carmalt. 2006. "Environmental Psychology and Sustainability in High-Rise Structures." *Technology in Society* 28 (1–2): 157–167. http://dx.doi.org/10.1016/j.techsoc.2005.10.016 [2]Nieuwenhuis, Marlon, Craig Knight, Tom Postmes, and S. Alexander Haslam. 2014. "The Relative Benefits of Green Versus Lean Office Space: Three Field Experiments." *Journal of Experimental Psychology: Applied* 20 (3): 199–214. https://doi.org/10.1037/xap0000024
	Increased healing rates	X[1]			[1]U.S.A.	[1]Ulrich, Roger S. 1984. "View through a Window May Influence Recovery from Surgery." *Science* 224 (4647): 420–421. https://doi.org/10.2307/1692984
Fractal patterns						
	Aesthetically pleasing	X[1]			[1]U.S.A.	[1]Taylor, Richard P., Branka Spehar, Paul Van Donkelaar, and Caroline M. Hagerhall. 2011. "Perceptual and Physiological Responses to Jackson Pollock's Fractals." *Frontiers in human neuroscience* 5: 60. http://dx.doi.org/10.3389/fnhum.2011.00060
	Stress reduction	X[1]			[1]U.S.A.	[1]Taylor, R. P. 2006. "Reduction of Physiological Stress Using Fractal Art and Architecture." *Leonardo* 39 (3): 245–251

Table A.2 Researched human–nature benefits

	Primary benefit	Country of study	References
	Increased well-being		
Secondary Benefits	*Short-term stress recovery* *Mental fatigue recovery* *Faster recuperation from illness* *Increased liveability*	Norway	Velarde, Ma D., G. Fry, and M. Tveit. 2007. "Health Effects of Viewing Landscapes – Landscape Types in Environmental Psychology." *Urban Forestry & Urban Greening* 6 (4): 199–212. https://doi.org/10.1016/j.ufug.2007.07.001
	Stress reduction *Health promotion* *Rehabilitation* *Disease prevention*	Japan	Park, BumJin, Yuko Tsunetsugu, Tamami Kasetani, Takahide Kagawa, and Yoshifumi Miyazaki. 2010. "The Physiological Effects of Shinrin-Yoku (Taking in the Forest Atmosphere or Forest Bathing): Evidence from Field Experiments in 24 Forests across Japan." *Environmental Health and Preventive Medicine* 15 (1): 18–26. https://doi.org/10.1007/s12199-009-0086-9
	Decreased blood pressure	Japan	Li, Q., T. Otsuka, M. Kobayashi, Y. Wakayama, H. Inagaki, M. Katsumata, Y. Hirata et al. 2011. "Acute Effects of Walking in Forest Environments on Cardiovascular and Metabolic Parameters." *European Journal of Applied Physiology* 111 (11): 2845–2853. https://doi.org/10.1007/s00421-011-1918-z
	Increased activeness in natural environment	U.K.	Bowler, Diana, Lisette Buyung-Ali, Teri Knight, and Andrew Pullin. 2010. "A Systematic Review of Evidence for the Added Benefits to Health of Exposure to Natural Environments." *BMC Public Health* 10 (1): 456. http://www.biomedcentral.com/1471-2458/10/456
	Increased well-being from: • *Viewing nature* • *Being in a natural environment*	Australia	Pryor, Anita, Mardie Townsend, Cecily Maller, and Karen Field. 2006. "Health and Well-Being Naturally: 'Contact with Nature' in Health Promotion for Targeted Individuals, Communities and Populations." *Health promotion journal of Australia : official journal of Australian Association of Health Promotion Professionals* 17 (2): 114–123. http://search.proquest.com/docview/68771442?accountid=10382
	Stress reduction *Pain reduction*	U.S.A.	Ulrich, Roger S. 2006. "Essay: Evidence-Based Health-Care Architecture." *The Lancet* 368: S38–S39. http://search.proquest.com/docview/199067283?accountid=10382
	Reduction in indoor air pollution *Stress reduction* *Psychological benefits*	U.S.A.	Wener, Richard, and Hannah Carmalt. 2006. "Environmental Psychology and Sustainability in High-Rise Structures." *Technology in Society* 28 (1–2): 157–167. http://dx.doi.org/10.1016/j.techsoc.2005.10.016
	Reduction in indoor air pollution *Odour reduction*	Japan	Oyabu, Takashi, Ayako Sawada, Takeshi Onodera, Kozaburo Takenaka, and Bill Wolverton. 2003. "Characteristics of Potted Plants for Removing Offensive Odors." *Sensors and Actuators B: Chemical* 89 (1–2): 131–136. http://dx.doi.org/10.1016/S0925-4005(02)00454-9
	Reduction in indoor air pollution	Portugal	Pegas, P. N., C. A. Alves, T. Nunes, E. F. Bate-Epey, M. Evtyugina, and C. A. Pio. 2012. "Could Houseplants Improve Indoor Air Quality in Schools?" *Journal of Toxicology and Environmental Health, Part A* 75 (22–23): 1371–1380. https://doi.org/10.1080/15287394.2012.721169
	Attention restoration		
	Perceived restorativeness	Sweden	Tenngart Ivarsson, Carina, and Caroline M. Hagerhall. 2008. "The Perceived Restorativeness of Gardens—Assessing the Restorativeness of a Mixed Built and Natural Scene Type." *Urban Forestry & Urban Greening* 7 (2): 107–118. http://dx.doi.org/10.1016/j.ufug.2008.01.001
	Stress reduction	U.S.A.	Kaplan, Stephen. 1995. "The Restorative Benefits of Nature: Toward an Integrative Framework." *Journal of Environmental Psychology* 15 (3): 169–182. http://dx.doi.org/10.1016/0272-4944(95)90001-2

(continued)

Table A.2 (continued)

	Primary benefit	Country of study	References
	Stress reduction		
		[1]Japan, [2]Finland	[1]Tsunetsugu, Yuko, Juyoung Lee, Bum-Jin Park, Liisa Tyrväinen, Takahide Kagawa, and Yoshifumi Miyazaki. 2013. "Physiological and Psychological Effects of Viewing Urban Forest Landscapes Assessed by Multiple Measurements." *Landscape and Urban Planning* 113 (0): 90–93. http://dx.doi.org/10.1016/j.landurbplan.2013.01.014 [2]Tyrväinen, Liisa, Ann Ojala, Kalevi Korpela, Timo Lanki, Yuko Tsunetsugu, and Takahide Kagawa. 2014. "The Influence of Urban Green Environments on Stress Relief Measures: A Field Experiment." *Journal of Environmental Psychology* 38 (0): 1–9. http://dx.doi.org/10.1016/j.jenvp.2013.12.005
	Aesthetically pleasing *Architectural implications*	U.S.A.	Taylor, R. P. 2006. "Reduction of Physiological Stress Using Fractal Art and Architecture." *Leonardo* 39 (3): 245–251
	Relaxation	Japan	Ikei, Harumi, Misako Komatsu, Chorong Song, Eri Himoro, and Yoshifumi Miyazaki. 2014. "The Physiological and Psychological Relaxing Effects of Viewing Rose Flowers in Office Workers." *Journal of Physiological Anthropology* 33 (1): 6. http://www.jphysiolanthropol.com/content/33/1/6
	Decreased violence and crime		
	Aesthetically pleasing *Physiological effects*	U.S.A.	Kuo, F. E., and W. C. Sullivan. 2001. "Environment and Crime in the Inner City - Does Vegetation Reduce Crime?" *Environ. Behav.* 33 (3): 343–367
	Aesthetically pleasing		
	Attention restorative *Stress reduction*	U.K.	White, Emma V., and Birgitta Gatersleben. 2011. "Greenery on Residential Buildings: Does It Affect Preferences and Perceptions of Beauty?" *Journal of Environmental Psychology* 31 (1): 89–98. http://dx.doi.org/10.1016/j.jenvp.2010.11.002
		[1]U.S.A., [2]U.S.A	Taylor, R. P., B. Spehar, P. van Donkelaar, and C. M. Hagerhall. 2011. "Perceptual and Physiological Responses to Jackson Pollock's Fractals." Biederman, Irving, and Edward A. Vessel. 2006. "Perceptual Pleasure and the Brain: A Novel Theory Explains Why the Brain Craves Information and Seeks It through the Senses." *American Scientist* 94 (3): 247–253. https://doi.org/10.2307/27858773
	Improved mental health		
	Depression recovery *Cognitive and memory increase*	U.S.A.	Berman, Marc G., Ethan Kross, Katherine M. Krpan, Mary K. Askren, Aleah Burson, Patricia J. Deldin, Stephen Kaplan, Lindsey Sherdell, Ian H. Gotlib, and John Jonides. 2012. "Interacting with Nature Improves Cognition and Affect for Individuals with Depression." *Journal of Affective Disorders* 140 (3): 300–305. http://dx.doi.org/10.1016/j.jad.2012.03.012
	Cognitive enhancement		
	Attention restoration *Therapeutic benefits*	U.S.A.	Berman, Marc G., John Jonides, and Stephen Kaplan. 2008. "The Cognitive Benefits of Interacting with Nature." *Psychological Science* 19 (12): 1207–1212. https://doi.org/10.2307/40064866
	Increased healing rate		
	Stress reduction	South Korea	Park, S. H., and R. H. Mattson. 2008. "Effects of Flowering and Foliage Plants in Hospital Rooms on Patients Recovering from Abdominal Surgery." *HortTechnology* 18 (4): 563–568
	Stress reduction *Therapeutic benefits*	U.S.A.	Ulrich, Roger S. 1984. "View through a Window May Influence Recovery from Surgery." *Science* 224 (4647): 420–421. https://doi.org/10.2307/1692984

(continued)

Table A.2 (continued)

	Primary benefit	Country of study	References
	Increased productivity		
	Increased positive mood	Australia	Daly, John, Margaret Burchett, and Fraser Torpy. 2010. "Plants in the Classroom Can Improve Student Performance." *National Interior Plantscape Association*
	Increased air quality *Increased concentration* *Increased positive mood*	Netherland, U.K.	Nieuwenhuis, Marlon, Craig Knight, Tom Postmes, and S. Alexander Haslam. 2014. "The Relative Benefits of Green Versus Lean Office Space: Three Field Experiments." *Journal of Experimental Psychology: Applied* 20 (3): 199–214. https://doi.org/10.1037/xap0000024

Appendix B
Google Trends

The graphs are broad and purely indicative, but interesting in revealing trends. The drop-off at the end of some is only due to the lack of inputted data at that point. The high peaks in biophilia and biophilia hypothesis coincide with the release of a music album by popular Icelandic songstress, Björk.

Biophilia:

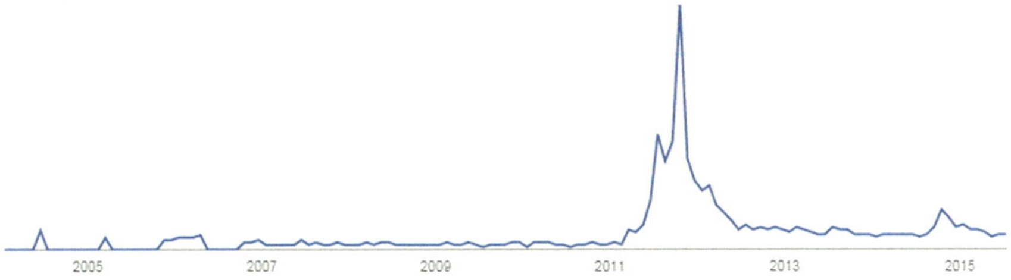

Fig. B.1 Biophilia Google Trends

Biophilia Hypothesis:

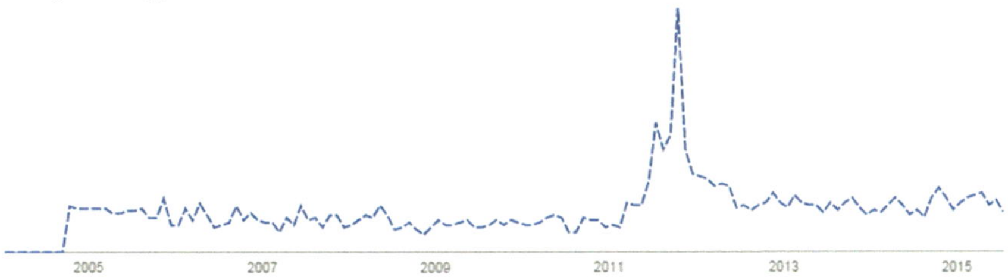

Fig. B.2 Biophilia hypothesis Google Trends

J. Söderlund, *The Emergence of Biophilic Design*, Cities and Nature,
https://doi.org/10.1007/978-3-030-29813-5

Fig. B.3 What countries are looking at Biophilia Hypothesis

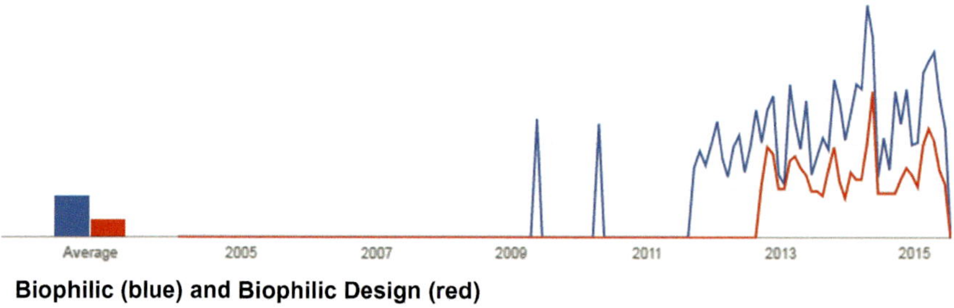

Fig. B.4 Biophilic and biophilic design Google Trends

See Fig. B.1.
See Figs. B.2, B.3 and B.4.

GPSR Compliance

The European Union's (EU) General Product Safety Regulation (GPSR) is a set of rules that requires consumer products to be safe and our obligations to ensure this.

If you have any concerns about our products, you can contact us on ProductSafety@springernature.com

In case Publisher is established outside the EU, the EU authorized representative is:

Springer Nature Customer Service Center GmbH
Europaplatz 3
69115 Heidelberg, Germany

Batch number: 10756376

Printed by Printforce, the Netherlands